ELEMENTS OF WAR

THEORY OF KNOWLEDGE, AND WAR

GRAY RINEHART

STORMWATCH

Praise for Gray Rinehart's novel, *Walking on the Sea of Clouds*:

"Two things are immediately clear. First, Gray Rinehart knows his field(s) inside out; and second, he writes with grace, skill, and professional polish. What more could any reader ask?"

—Mike Resnick, multiple Hugo-award-winning author

"There is a very rare and special pleasure that comes from reading a beautifully written book from a true expert in his field.

"This book will be treasured by anyone who has ever dreamt of visiting the Moon, walking on another world, or bathing beneath the light of a distant star."

—David Farland, author of the NYT-bestselling *Runelords* novels

"Much like *The Martian, Walking on the Sea of Clouds* puts you on a lifeless rock and makes you think about why we explore new frontiers even as it explains how it can be done."

—*Booklist Online*

Praise for *Quality Education* by Gray Rinehart:

"*Quality Education* . . . is a well-written combination of history, research, strategy, and challenge for educational reform [and] everyone who is willing to step up to the challenge of reforming schools should read it.

—Susan Hillingmeyer, *Quality Progress*

"*Quality Education* looks at education as a complete system that ties together all segments. It fosters a truly collaborative vision that brings students and adults together in a powerful partnership. . . . I heartily support [Rinehart's] inclusion of students as actual contributors in education reform."

—Gary Goldman, Quality Improvement Associates, Chicago, Illinois

ALSO BY GRAY RINEHART

Nonfiction:

Quality Education: Why It Matters, and How to Structure the System to Sustain It

Fiction:

Walking on the Sea of Clouds

Music:

Truths and Lies and Make-Believe

Distorted Vision

Taking You Out to See the Stars

ELEMENTS OF WAR

ELEMENTS OF WAR

THEORY OF KNOWLEDGE, AND WAR

GRAY RINEHART

STORMWATCH

Elements of War: Theory of Knowledge, and War

Published by Stormwatch, Cary NC
Cover design by the Art and Design of Christopher Rinehart
Interior design by Drew Avera

Library of Congress Control Number:2019912085

Print ISBN: 978-0-9982092-3-4
EPUB ISBN: 978-0-9982092-4-1
Kindle ISBN: 978-0-9982092-5-8

ELEMENT:

One of the simplest or essential parts or principles of which anything consists, or upon which the constitution or fundamental powers of anything are based.
(*Webster's Revised Unabridged Dictionary*)

a constituent part: as
a *plural* **:** the simplest principles of a subject of study
(Merriam-Webster On-Line)

WAR:

A contest between nations or states, carried on by force, whether for defence, for revenging insults and redressing wrongs, for the extension of commerce, for the acquisition of territory, for obtaining and establishing the superiority and dominion of one over the other, or for any other purpose; armed conflict of sovereign powers; declared and open hostilities.
(*Webster's Revised Unabridged Dictionary*)

a state of usually open and declared armed hostile conflict between states or nations
(Merriam-Webster On-Line)

For all who serve

CONTENTS

FOREWORD

If one accepts the maxim that chance favors the prepared mind, then it is equally axiomatic that the most prepared mind will enjoy the greatest probability of success. While obviously important to any human endeavor, maximizing the probability of success is never more vital than in the conduct of warfare. This is simply because the failure of a warrior or group of warriors to adequately prepare for combat risks the lives of not only the individual combatant or unit, but the entire campaign in which they participate. This in turn can jeopardize the nation, and in today's combat environment might very well risk worldwide catastrophe. It is therefore not only a necessity but an obligation for every individual pursuing the vocation of arms to prepare as completely as possible for the fulfillment of their military duties.

On one level of abstraction, this preparation is encompassed by maintaining an acceptable minimum level of technical and tactical proficiency. Such proficiency allows for the maintenance of combat readiness, and the attendant logistical necessities of the military establishment during peacetime. At the same time however, the military establishment must be prepared to defend the national interests, including those of its allies. It is in this arena that military preparation becomes problematic, due to the number of unknowns and imponderables involved.

The national or strategic goals of an enemy may not be immediately apparent; neither is the degree of an enemy's willingness to pursue those goals, or the means at their disposal. Alignment and alliances are similarly not always obvious, and are always subject to change. The European "Wars of Empire" and the World Wars of the Twentieth Century clearly illustrate that the nature of conflicts can change in the macrocosmic sense, as well as in the microcosmic environment of the battlefield. There exists an increasing probability of a paradigm-shifting conflict at the scale of the incredibly vast—Space Warfare—where the current Administration's proposal of a "Space Force" is certain to cause no end of logistical and administrative issues; or the incredibly small—the ones and zeroes of binary computer code—where a "virtual" military presence can have a disproportionate impact on an enemy (see Stuxnet). The role played by technology only aggravates and accelerates this process. For example, the evolution of gunpowder weaponry, a technology at its fullest possible development, can be easily studied at an academic distance. Conversely, the continuing and accelerating development of computer technology required the expansion of the "battlefield" into the "battlespace," in which the "electromagnetic spectrum and the information environment" are equally vital, and equally dangerous landscapes that must be traversed. It is therefore reasonable for the military professional, and policy makers for that matter, to make every possible attempt to determine what is known about an enemy, thereby limiting what is unknown.

Many military professionals will confidently dismiss any discussion of such matters as sterile conjecture, retreating into the language of catchphrases and maxims, concluding that success in warfare is no more complicated than putting "steel on target" or "getting there firstest with the mostest." They will likely rest confident in the belief that the nature of warfare does not change, and that the maxims they believe in will hold them in good stead during any conflict now or in the future. Conversely, many, if not most academics tend to minimize the importance of actual military operations as being of less significance within the greater construct of historical causality. It turns out that both are correct, but only in a very limited sense.

While it is true that the nature of war does not change, the means by which they are fought, and their underlying purposes can change virtually overnight. Similarly, while all history is not military history, and wars are, as Professor Michael Howard quite rightly asserted, conflicts between societies, the predicates of those societies, and their impact on military operations have direct and immediate impact on military personnel at every level of conflict. Recent American history illustrates the consequences of the US military establishment bringing the most technologically advanced armed force in history to bear against societies in Asia with fundamentally different social, military and strategic predicates. The American experience in Vietnam clearly demonstrated that knowing the nature of the enemy is equally as important as the ability to defeat him on the battlefield. Even more recent events suggest that neither the old Soviet Union, nor the last four U.S. Presidential administrations possessed an adequate understanding of the multi-ethnic, multi-lingual environment of modern Afghanistan. Knowledge therefore becomes an indispensable component of military preparation—thereby reducing as much as possible the "Fog and Friction" that so captivated von Clausewitz nearly two centuries ago.

This requirement for knowledge of an opposing force, their strategic and/or national goals, their willingness to pursue them and the underlying reasons for those goals, justifies and indeed mandates the maintenance of a "military intelligence" apparatus allowing for the most complete possible understanding of enemy dispositions. This is the first requirement in countering his tactics and defeating his strategy. However much information we may develop on a given opponent, and whatever data-based analysis arises from that information, we still must deal with the reality that we cannot *know* what the enemy will do until he does it. In this domain maxims are not viable, but probabilities certainly are. It is therefore essential for theory to be developed and studied—for conflicts mandate the development of explanatory theory, in turn providing greater insight into the *probable* nature of future conflicts. Military Intelligence, when taken to the greatest possible extreme will provide analysis of the enemy at present and the most probable disposition of his forces in the future. Admittedly such predictions can be made without theory, and in fact can be made with or without adequate information—the difference being that between technical and tactical proficiency, and the work of science fiction.

Only through the development and study of theory can we determine with any degree of probability what a future enemy might do. It might well be the case that attempting to develop a viable theory of knowledge and war demonstrates our reach exceeding our grasp—but it remains clear that the attempt increases our corporate understanding of the underlying processes, thereby providing greater insight into the probabilities. The human variables are unknown and largely unknowable, so maxims *per se* are impossible. A century ago Hans Delbruck concluded that the only true maxim of war is that there are no enduring maxims. This is likely the case, but analysis of past performance within a structured framework should result in reasonable expectations and assumptions on which strategic planning is based. That analysis, and the underlying theory thereof is the right and proper intellectual pursuit of the military professional, both now and in the future.

James Galt-Brown, PhD
Tifton, Georgia
August 2019

"For now we see through a glass, darkly . . ."
— 1st Corinthians 13:12 (KJV)

PREFACE

It can be dangerous to think of war in the abstract, especially for those who have the awesome and awful responsibility of making war. Unfortunately, it is all too easy to think of war as an abstraction, and to speculate about or analyze it without taking full enough account of the concrete realities of fire and blood and pain. It is all too easy for those of us who served in support roles in the rear echelon to produce volumes like this one, which questions—perhaps in ignorance, perhaps in arrogance—some deeply-held "truths" about war and warfare.

Why might we question such truths, such accepted wisdom? It may be better to ask why we don't question them more often.

Everyone is familiar with the experience of learning something, believing it to be true, and finding out later it was not quite accurate. Perhaps the difference was in the details—learning that the planet has two north poles, geographic and magnetic, for instance—or perhaps what we learned was false or incomplete, e.g., the characteristics of life at one or both of those poles. We "learned" that the moon was made of green cheese, that Mars had canals, and that the solar system had nine planets; one was nonsense, one supposition, and the third science; but, after the discovery of the tenth planet (at the time "2003 UB313," now Eris) and then the rejection of it and Pluto as true planets in favor of the "dwarf planet" designation, we now know that all three things we learned were wrong—or, in the last case, premature.

This process—collecting new information, enjoying or enduring new experiences, and reevaluating what we learned—can be uncomfortable, so we may not appreciate it at the time. We may think of it as going through intellectual and emotional growing pains. But when it comes to history, this growth experience can produce mistrust if we put too much stock in what we already learned. We may deride new interpretations as "revisionist history," forgetting that *all* history must be subject to revision —literally, "looking again"—as new facts are discovered.

Unfortunately, facts are not always recognizable or readily available. Where facts are obscure or absent, we must interpret, interpolate, and speculate in order to derive anything approaching understanding or discernment. This is a natural process, i.e., inherent to our nature as thinking beings, and we routinely accept an abridged understanding of things that cannot be proved by fact or rationale.

In the field of military thought as in the rest of life, many of us were taught things that upon closer examination turned out to be, if not exactly wrong, at least *less right* than we were led to believe. "Led to believe" is a key term, especially as we are considering military matters. Military leaders enjoy high stature, and their pronouncements carry automatic authority. Furthermore, few military personnel can afford the time to dig into all the potential explanations of an event to validate or verify them. Claims that seem self-evident become part of the body of "truth" that is accepted and taught, accepted and taught, with little question and less criticism.

But some of those "truths" may be less reliable than we have come to believe, which is why history as a collection of facts—the "what"—may be interesting, but history as a collection of explanations—the "why"—is much more fascinating even if much more difficult. Sometimes the story of "why" is written in chronicles or memoirs that tap directly into the motivations of the main players, but that, unfortunately, is rare. Even where first-hand accounts are available, they limit the explanations to what the players knew. What they knew could be wrong—and certainly was incomplete. And what are we to do when what is known (or knowable) shows some of our "truths" to be less dependable than we might like to think?

As something of a military epistemology, this book examines certain "truths" in the link between knowledge and war; it considers the relationship between *theory* (i.e., methods of understanding the world, as in Theory of Knowledge) and *practice* (i.e., methods of prosecuting) with respect to war. You will have to judge whether this single volume does justice to such a broad topic, but the book itself grew from a more humble beginning. Much like an old city that grew up from a trading post or port, in which modern thoroughfares follow what were once cart paths, over time entire sections have been razed or renovated such that the resulting landscape is hardly recognizable compared to its origin. This book began as a single unpublished article, grew into a three-article series that also was never published, then shrank back again to a single article published on-line in 1998 as "Theory of Knowledge, and War—First Steps Toward a Unified Theory" in *Air Chronicles Contributors' Corner* (the on-line journal of the US Air Force). Over the ensuing twenty years, it expanded in fits and starts into its present form.

In this book we will begin by considering the nature of war, and especially the risk of war, and proceed from there into the nature of theory and what theory has to contribute to our understanding of war. We will discuss several ways of describing war and warfare and whether they form coherent and useful theories; we will present some new ways of thinking about those descriptions of war and apply concepts from other fields to the topic; and we will derive what appear to be fundamental elements of war and warfighting, which we describe as Command and Momentum, driven by Will. We will test the elements of Command and Momentum against historical battles and conjectured future battles to see whether they hold up. Finally, we will revisit the topic of the risk of war, because understanding war involves accepting not only the risks associated with battle but the risk that battle will be necessary in the first place.

Sometimes finding out that what we thought we knew was not quite correct—that "we do not know what we think we know"—is more than uncomfortable: sometimes it shocks us, and a shock like that may lead us to avoid seeking deeper understanding. When we question those things we thought were true, we naturally question the authority figures—parents, teachers, other leaders—from whom we learned them. If we are not careful, our confidence in our forebears may dwindle too far. Our final response after such examination, though, is ours alone: whether we will strengthen our belief in the old truths because the counter arguments were not convincing, or whether we will abandon or modify those old beliefs in light of new evidence or argument.

In the end, let us hope together that this thinking of war in the abstract may benefit those who must fight wars in the real world.

Graham W. "Gray" Rinehart
Lieutenant Colonel, USAF, Retired
Cary, North Carolina
August 2019

1

WAR AND THEORY

Every time someone declares why certain battles and wars were won or lost—whether that person is a veteran, an observer, or a historian—and especially when someone declares how battles and wars may be fought and won in the future, the skeptic is privileged to ask, "How do we *know*?" Inflection is important in the question; this is not the same as asking, "*How* do we know?"

"*How* do we know?" is a question of process. *How* do we know that fatigue and dwindling strength, rather than treachery—or vice-versa—doomed the defenders at Thermopylae? *How* do we know that airpower was or was not decisive in Kosovo in the late 1990s, and that it would or would not have been decisive in any other campaign? *How* do we know . . . ? These are questions of determining cause and effect—or at least of suggesting cause and effect. These are the questions of history, and of doctrine.

But when we ask, "How do we *know*?", we are asking a different question. What is the character of these data, these evaluations, these inferences that lead us to *know*, as opposed to merely thinking or guessing? What does it mean to *know*? These questions bear more directly on the subject of this book—they get at the root of knowledge, and pare away the outer shell of "conventional wisdom." And, as we shall see, "To know in the sense of familiarity and to comprehend in clear ideas are . . . quite different matters."[1]

Likewise, whenever someone compares, say, Jomini and Mahan, or Clausewitz and Sun Tzu,[2] and recommends selecting one's approach or outlook over another's, the skeptic rightfully may ask, "Why?" Some may ask these questions just to annoy, others to play Devil's Advocate. Better still to ask questions like these out of an honest attempt to establish knowledge *qua* knowledge as it relates to war and warfare; i.e., to understand what we *know* as opposed to what we *think we know*.

Our understanding of military affairs is the legacy of many great minds and many great warriors. They offer us histories and analyses of wars and battles too numerous to count—both the battles and the analyses. The history of war is rife with examples of leaders who did not learn from it, and leaders who defied it, but in this age that history has grown so long that few of us—including this author—have read more than a speck of it. Thus we sometimes find it difficult to establish a common frame of reference for understanding war and warfare.

However, an overabundance of history is not the only problem we encounter when studying war.

We also encounter the problem that theory itself is not that highly esteemed in the military; historically, it has even been held in contempt.[3] Martial thinkers generally disdain theorizing about war, preferring to study the examples of experience, even (or especially) the vicarious experience of history. But experience itself can be a poor teacher, as observed by no less an eminent strategist than Sir Basil H. Liddell Hart. He wrote, "Direct experience is inherently too limited to form an adequate foundation either for theory *or for application*. At the best it produces an atmosphere that is of value in drying and hardening the structure of thought."[4] Recent US Air Force Doctrine holds a similar view of experience, but does not go so far as to link experience and theory. "While experience plays a major role in doctrine formulation, too great a reliance on past experience leaves one open to always fighting the last war. Experience must be tempered with current realities to develop future plans."[5]

How can we show that theory is neglected in military circles? For one thing, references to it in US Joint Doctrine are relatively scarce. Even though chapter 1 of *Joint Publication 1, Doctrine for the Armed Forces of the United States*, is entitled "Theory and Foundations," the word "theory" appears only once, when the text says that Prussian military theorist Carl von Clausewitz "believed that war is a subset of the larger theory of conflict."[6] It is likewise mentioned only once in another major doctrine document, which holds that "Joint operations doctrine is built on warfighting philosophy and theory derived from experience."[7] A good portion of *this* book will examine the relationship between theory and experience, but it seems clear that Joint Doctrine does not provide a ringing endorsement for the development or study of theory. While the old Army *Field Manual 1* mentioned theory once in its explanation of the purpose of doctrine—"Doctrine links theory, history, experimentation, and practice"—its successor, *Army Doctrine Publication 1*, mentions it not at all.[8]

The Air Force, for its part, seems to treat theory a bit better. It was not always so: before it was supplanted by the 1997 and later versions of *Air Force Basic Doctrine*, the old Air Force Manual 1-1, *Basic Aerospace Doctrine*, stated categorically, "Theory, of course, is no substitute for military genius or even for professional judgment."[9] Such a flat statement about the usefulness (or uselessness) of theory—that theory is no substitute for genius—is woefully inadequate, especially if military genius is unavailable. What of the possible relationship between genius and theory, or judgment and theory? Presumably the doctrine writers thought no such relationships were worth examining. The implication was that once one has obtained some level of professional judgment, even if it does not reach the level of military genius, theory may be discarded as superfluous if not actually useless. In addition, this was stated "of course", implying it could be no other way, indeed implying that anyone who would dare extol the virtues of theory must be either pedantic or idiotic. (This author may qualify on both counts.) Despite the rhetoric, however, we shall see that the statement hides the full relationship between theory and the development of knowledge. We may be thankful that later iterations acknowledged a role for theory in military thought: "Theory is . . . designed to encourage debate and introspection with an eye towards improving military advantage. It is part of a vital, iterative investigation of what works under particular circumstances, and why."[10] (Interestingly, however, current Air Force doctrine still applies a "no substitute" sentiment with respect to the Principles of War: "Valid principles, despite how deeply they are held, are no substitute for sound, professional judgment . . . ."[11])

But the Marine Corps, among the services that make their doctrine public,[12] treats theory with the most respect. *Marine Corps Doctrine Publication 1, Warfighting*, devotes an entire chapter to the theory of war, enjoins every Marine "to be a student of the art and science of war," and notes that Marine "officers especially are expected to have a solid foundation in military theory and a knowledge of military history and the timeless lessons to be gained from it."[13] *MCDP 1* also states categorically that "a leader without either interest in or knowledge of the history and theory of warfare—the intellectual content of the military profession—is a leader in appearance only."[14]

Perhaps it should not surprise us that the Marines hold theory in higher regard than other services. If we think of the battlefield, the traditional arena of military endeavor, we may conceive of a helpful metaphor for the role of theory. The lay of the land and avenues of approach, along with prevailing and expected weather conditions, produce a situation that the commanders of the facing armies will each try to turn to their own advantage. The battlefield is the given, the physical reality with which the commanders must contend. How they decide to use the battlefield—placement of forces, timing of feints and maneuvers, and so forth—will be based on those prevailing conditions as interpreted with the aid of previous experience (including training) and the commander's prediction—his *theory*—of how the battle is likely to turn.[15] Without that final element, without theory or the ability to theorize, the commander will be unable to act intelligently or decisively.

As much as some military personnel may wish to ignore theory, the truth is that theory is necessary to understanding and successfully fighting any war. To answer the question of the possible relationship between genius, or judgment, and theory: theory is not the spark of inspiration, but the flint with which the spark may be struck from the steel of experience and thought. Theory is the foundation for genius and the framework for professional judgment.

2

WAR IN THE ABSTRACT

To say with General William Tecumseh Sherman that, "War is hell,"[1] while descriptive of the *experience* of harsh conditions and fierce, prolonged fighting, does not help us understand what constitutes war and warfare. The statement itself is true enough, and whether it is pithy, colorful, or flippant is a matter of opinion, but it does not tell us what war itself is. Nor is any similar quote likely to do so.

A statement such as "War no longer exists," made by General Sir Rupert Smith in his 2005 book *The Utility of Force*, is more problematic. General Smith maintained that

> Confrontation, conflict and combat undoubtedly exist all round the world . . . and states still have armed forces which they use as a symbol of power. None the less, war as cognitively known to most non-combatants, war as battle in a field between man and machinery, war as a massive deciding event in a dispute in international affairs: such war no longer exists.[2]

General Smith maintains that war as he describes it has changed into something else. Even if that is true, that new thing—which still involves armed conflict—is likely to continue being called "war." We hope we may be forgiven a certain amount of pessimism if we say that even his qualified statement will only hold true until the next war occurs; therefore, it behooves us to understand what war is and how to conduct it with the greatest chance of success.

We need a common understanding at even a basic level if we are to communicate with one another on the subject. For example, British historian John Keegan's June 1999 claim that "the capitulation of President Milosevic proved that a war can be won by airpower alone"[3] can only be true if we agree on two things: first, that airpower was the necessary and sufficient cause of hostilities' ceasing; and second, that the hostilities in Kosovo from March to June 1999 constituted a war. We will return to that example later; but for now, let us attempt to build a common understanding of war.

What is war? This single chapter will never capture the whole of a subject as complex as war and warfare—indeed, entire books leave key parts of the subject unexplored. This single chapter may not even define the term satisfactorily, but we hope it may provide a suitable beginning and a foundation for what follows.

Two turn-of-the-century studies of the causes of war—John Vasquez's *The War Puzzle* and Alexander Moseley's *A Philosophy of War*—each stressed the importance of defining "war" carefully to establish a firm basis for discussion. Vasquez cautioned that "poor definitions not only lead to confusion but often end up telling you more about the person who stipulated the definition than about the subject,"[4] and Moseley cautioned that "definitions are philosophically awkward beasts."[5] Vasquez and Moseley examined several definitions, the former as a basis for a scientific examination of wars' causes and the latter as a basis for the philosophical underpinnings of why humans wage war, something which military historian Martin van Creveld covered with admirable thoroughness in *The Transformation of War*.[6] However, we are concerned less in this book with the causes of war and more with its conduct; i.e., less with *why* humans wage war and more with *how* we can wage war most effectively.

Just as we must crawl, then walk, then run, we will begin with basics and work our way up from there. With that in mind, the simplest explanation of war may be "conflict writ large." As that is almost absurdly simple, what of more formal definitions?

In addition to the definitions posted in the front of this book, the Oxford English Dictionary defines war as "hostile contention by means of armed forces, carried on between nations, states, or rulers, or between parties in the same nation or state."[7] Furthermore, the OED defines "warfare" as "the action of carrying on, or engaging in, war; the act or state of conflict." Definitions such as these will have to suffice since the US Department of Defense has not adopted an official definition; there is no entry in the *Dictionary of Military and Associated Terms* for "war" or "warfare."[8] This is curious: either the DoD has accepted *a priori* that everyone understands what "war" is, so no definition is needed, or defining it adequately has proven too difficult.

Deciding that a definition is unnecessary seems an inappropriate conclusion; though he was not commenting on the lack of an official definition, in *The War Puzzle* John Vasquez stated that "Most . . . seem content to work with modified everyday or ordinary definitions, since 'we all know what a war is,' a sure sign that we may not know anything theoretically significant about war."[9] Nevertheless, to date we seem to have done well enough without an official definition, so perhaps one would be unhelpful; or perhaps, because we deal so often with different levels and types of conflicts, a single definition of war may hinder more than help.

Let us return, then, to the idea of conflict on different scales, beginning with the assumption that we are discussing violent conflict. A man vying against another man may be playing a game, or attempting some nonviolent persuasion, or involved in a deadly clash; of the three, within the context of this book we are concerned first and foremost with the violent option. Whether contested with bare fists or sticks or broken bottles or knives, we give this contest the generic term of a "fight." If we involve more people, we may call the result a "brawl" or even a "riot." Somewhere along the line of scale we encounter families (tribes or clans) fighting each other, which we may term a "feud" if not a "tribal war." Extra-family groups may also engage in conflict with one another. Often we associate these with political or ideological groups, but such conflicts may be economic or even criminal in nature; e.g., consider piracy during the Age of Exploration or off the African coast today, or the gangland wars in Chicago and other US cities, especially during the time of Prohibition. Gradually the conflicts become larger and larger until we come to the common notion of "war," that of polities fighting other polities. At this point the distinction blurs, as everything from the struggles of Greek city-states to the World Wars of the last century are all considered to be "war" in the classic sense.

This list of comparisons does not, however, include all the possible permutations; we may doubt any readable list really would. One variation that began receiving more attention in the mid-1990s would involve computers and networks being used to contest against other computers and networks in what came to be called "Net-war."[10] The war that most concerns us in the second decade of the 21st century, because it is the one in which we have been embroiled for some time, is predominantly characterized by small groups—i.e., sub-national organizations—fighting a nation or a group of nations. Depending on the limits of the battlefields involved, this is usually called civil war, guerrilla war, or terrorism. The advent of "global terrorism," as it came to be known, added a dimension to this conflict that did not exist prior to about 1990.

Before we continue, let us say a few words about the terminology of terrorism.

The word "terrorism" is very widely accepted, and in fact has grown to be somewhat overused, but it may not be the best word to describe the kind of violence that strikes at traditional noncombatants across national borders. In fact, the Department of Defense in mid-2005 tried to shape the debate when it dropped the name "Global War on Terrorism" in favor of the more unwieldy but more descriptive "Global Struggle Against Violent Extremism." That term was thought to cast a wider net, since it implied that all violent extremists are not terrorists and all struggles are not wars (more on this below). Whether "violent extremist" was a better term was debatable, but only for a short time; the better-known GWOT was re-adopted within days.

It is hard to pinpoint exactly when "terrorist" (whether theo-, politico-, eco/enviro-, or econo-) stopped meaning an internal agitator: i.e., a person or group with the aim of creating internal political change by the selective application of force—and the threat of further force—against their own population. Outside of occasional hostage taking, which usually involved citizens of other countries chosen for their propaganda value (e.g., the 1972 Munich Olympics, the 1985 hijacking of the *Achille Lauro*, etc.), "traditional" terrorism was largely an internal affair. The transition occurred gradually, and attention shifted away from such locally-focused groups as the Irish Republican Army (IRA), Palestinian Liberation Organization (PLO), Kosovo Liberation Army (KLA), etc., to cross-national groups typified by Al Qaeda.[11] The rise of the so-called Islamic State (whether in Syria [ISIS] or the Levant [ISIL]) represented a graduation of sorts for terrorists into, shall we say, more grown-up warfare.

As terrorism became a cross-border phenomenon and the terrorists sought larger prizes, the traditional (if that can be the correct word) aim of instilling terror in the populace seemed to be dropped. Within a nation, terror might keep a population from interfering with a coup, or coerce a population into accepting a new government. In combating a colonial presence, guerrilla action (which is not terrorism in the limited sense, although terrorist-style attacks are characteristic of it) may wear down the larger, slower occupying forces such that they either withdraw, are ultimately defeated through fatigue, or become so vicious to the population that the people join the guerrillas to oust the occupiers. With attacks against targets in separate, sovereign nations, however, by persons affiliated with outside non-national and non-nationalistic groups, it is difficult to see what political aims these new combatants wish to achieve; therefore, terrorism in the classic sense does not apply. "Low-intensity conflict" is more descriptive than "terrorism" but the latter has a better ring to it; however, in the search for a better word (e.g., "violent extremism") we may wish to separate this new brand of conflict from the others. Unfortunately, the terrorist label has been so broadly used that it may not easily be displaced.

We may propose any number of possible names. In this case the word "struggle" may not be the best choice—to English speakers it denotes a larger, lower intensity conflict than a war, but it translates into Arabic as "jihad." As terrorism has become something of a proto-war, it may be better to consider other names for war, such as some variation of words from which the English word "war" derived (e.g., werra). On the other hand, we might consider names for war from other languages (e.g., guerra, krieg, krig, oorlog). Even those, however, may lose something in the translation.

Let us, then, propose a new term. Those who seem determined to wage indiscriminate conflicts—wars in all but name—across national borders, often against civilian targets, and for reasons that have nothing to do with national sovereignty or national interests but rather sub-national and often ideological interests, may be considered *bellicose* in the strongest sense of the word: warlike. With that as the basis, we may term them "bellicosi" and characterize them as warlike without being warriors. Alternately, we might characterize them more as regular fighters and consider them *pugnacious*; we could then use that basis for a term (perhaps "pugnaces"), but that may not be any better. Outside the pages of this book, no alternative term may end up supplanting "terrorist," but "bellicosi" at least has the quality of being different enough that it does not connote the same kind of actions as those of "regular" terrorists.

Why this aside into the terminology of terrorism? One, because it illustrates some of the danger of thinking about war in the abstract: that we can spend more mental effort on fine points than those subjects really deserve. Two, because in the current age, the onslaught of globalization and the permeation of terrorism across borders has changed the nature of conflict.

The bellicosi do not seem loyal to political artifices like countries, but rather to identify with ideologies and smaller people groups. Few nation-on-nation conflicts are in motion today; instead, the world teems with more concentrated violence that combines the deep-seated hatred of clan fighting with the destructiveness of modern weapons.[12] The advent and continuation of this new type of conflict, which General Sir Rupert Smith called "war amongst the people," led him to say that traditional war "no longer exists."[13]

This new conflict is also characterized by the blurring of the line between combatants and noncombatants; indeed, we may be justified in asking whether there are any real "noncombatants" anymore. Certainly in modern industrial society it is not necessary to carry an actual weapon to participate in national defense activities, or to be affected by military action. Even those far removed from defense activities may suffer in wartime. "Collateral damage" is not a modern phenomenon, though modern warfare has exacerbated it; consider the naval blockade of Germany in World War One that resulted in the slow starvation of the country: ". . . in the last winter of the war nearly three-quarters of a million Germans were to die of hunger."[14] It may be said that those deaths, though caused by legitimate military action, were incidental and unintended; in today's world, however, one's very existence may be enough to be targeted, deliberately or otherwise, by the bellicosi.

We may debate whether this prolonged, intense conflict is a remnant of colonialism or a unique phenomenon, though such a debate is unlikely to resolve anything. In contrast to such abstract debate we must place the facts of this conflict which, like the facts of any other, are brutal; we cannot ignore them, we can only confront them.

3

THE RISK OF WAR

Why do we need to spend so much mental energy trying to understand war and warfare? In this supposedly enlightened age, some might forget that the "war to end all wars" did not succeed in doing so, and so make the mistake of believing that war is an outdated notion, a thing of the past, something that only happens in other places and to other people. They might read General Sir Rupert Smith's assertion that "war no longer exists" and read no further—and so never get to his exposition of a new type of warfare that is far more pervasive and possibly more dangerous than traditional war. This era is not the first in which many seem to believe that it is finally time to beat the swords into plowshares and the spears into pruning hooks.[1] Thankfully, those who believe such nonsense are usually not in charge of our protection; those protecting us know better.

We are far from reaching "the end of history,"[2] as Francis Fukuyama put it; and so, at the risk of sounding hawkish, or cynical about human nature, we can safely formulate this basic proposition: "men will fight." To expand the idea:

> men (inclusive of the human race, not specific to one gender)[3]
> will (it is unavoidable)
> fight (compete, clash, battle, struggle).

This is conceptually similar to what has become known as Pournelle's Axiom: "There will be war."[4]

Most readers of a book entitled *The Elements of War* presumably need no convincing that the proposition above and similar axioms are disappointingly accurate. Anyone to whom this is overly simplistic or hard to swallow may at least admit that it is hard to conceive a purely peaceful order of existence without holding in mind premises that are contrary to nature—all nature, not just human nature. For example, John Keegan writes, "It is difficult to see how the necessary system of environmental control could be carried on by creatures which lacked aggressive responses altogether."[5] Even "peace protesters" often express themselves through violence and mayhem, as the mob mentality builds among them; if they are not exactly warlike, they are still often less than pacifist.[6]

Furthermore, consider the various utopian fantasies we find throughout history and literature. Communism, for example, was supposed to represent an "evolution" of mankind that was to be achieved by "revolution" (itself a kind of economic and political "natural selection"—a violent enough process in nature as few species go gently into the good night); it was supposed to produce the idyll of the workers' paradise, but instead involved millions in toil and servitude to the State with only deprivation and deplorable conditions as their rewards. We may dream of idylls, but it seems humankind is not made for them. As the US Air Force's *Basic Doctrine* puts it,

> The modern Western tendency to view war as an aberration in human affairs, only occasionally necessary as an operation with limited aims or an all-out campaign to destroy a clearly recognized evil, often distorts our understanding of warfare and its purposes. Warfare is ingrained in the very nature of certain cultures. While for nation states, war is an instrument of policy aiming at political objectives; it is also, even within this context, a phenomenon involving the full range of human emotions and irrationalities. War has a dynamic of its own, often fueled by anger, fear, revenge, and hatred. Thus, the resort to violence rarely remains for long tied to cold, clear political objectives; it can—and has—moved in unexpected directions.[7]

Accepting "men will fight" as a basic prediction leads to the further proposition that "men will fight wars." Wars of some type, at some level of conflict.

Are some aims achievable only by conflict? Or, more to the point, are there collective, social, or national aims that can only be achieved by the application or the threat of violence? Optimism, or perhaps idealism, would suggest—no, would *hope*—not. The problem with idealism in this case is that violence is a two-party affair, and we may find ourselves embroiled in violence even if we do not wish it and try actively to avoid it. If the other party believes its aims are best achieved by violence—and we cannot help but think of the bellicosi and September 2001—we almost certainly will not dissuade them unless (and perhaps not even if) we pose a credible threat ourselves. Realism, then, if not actual pessimism, requires that we prepare for violence. Pragmatism, in turn, may lead us to rely on violence to achieve certain aims.

What are we to do with this prediction that some type of conflict is inevitable? To evaluate what type of conflict, we need to understand the risk of conflict. If, as suggested by John Keegan, the severity of war is increasing with time, and its probability is decreasing with time, then the risk of war—or at least the risk of military engagement, taken as the probability times the severity, as in risk management and hazard prevention—turns out to be fairly constant and relatively moderate, as shown in Figure 1.[8]

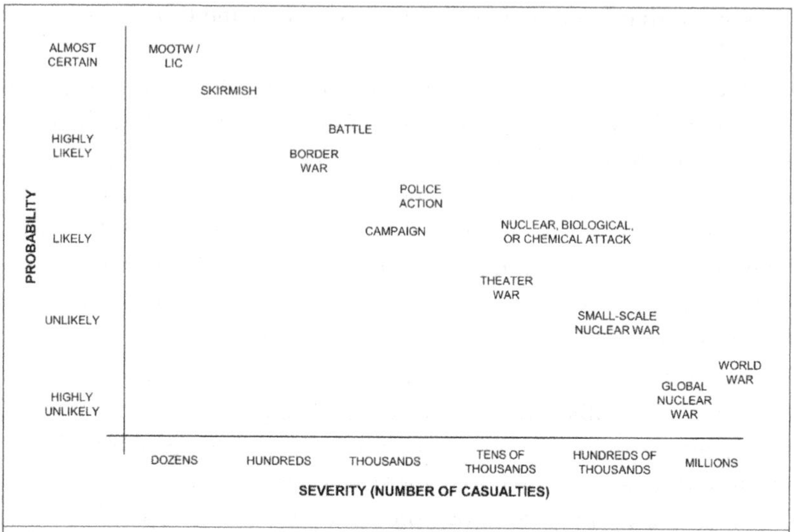

Figure 1. The risk of war in today's world. Exact placement on the chart may vary with different circumstances, but the overall risk (probability times severity) remains fairly constant. Note that expressing severity differently (as in number of civilians displaced, or economic damage) may yield different results.

A quick glance at some examples seems to confirm this. Low-intensity conflicts, skirmishes, and border wars—including the NATO actions in Kosovo (see below) and Libya—which may include irregular warfare and such things as the "war on drugs": high probability, but generally low severity of specific engagements even if the long-term severity of protracted conflicts can rise to staggering amounts. Martin van Creveld wrote in 1991 that, "Besides being numerically predominant," i.e., occurring frequently enough that we may assign them a high future probability, low-intensity conflicts "have also been far more bloody than any other kind of war fought since 1945," with "the total number of those who died . . . put at 20 million or more."[9] A few casualties at a time, in frequent (if not nearly constant) small-scale actions—even when they do not rise to the level of what we consider "war"—add up over time to horrendous totals.

How probable are these smaller conflicts? Even if we base our estimate only on those in which our own forces have fought, the answer may be surprising. A 2004 study tabulated 180 US Marine landings on foreign shores between 1800 and 1934, or "more than one per year. During roughly the same period, the U.S. Army deployed numerous small contingents in actions virtually all over the globe. Likewise, the

U.S. Navy, though small, was involved in many actions at sea over the same time span. . . ."[10] Marines, Soldiers, Sailors, and Airmen could relate to that level of operations tempo throughout the latter years of the 20th century, during which they were involved in high numbers of relatively small engagements; and that was even before multiple long deployments became the order of the day during the campaigns in and occupations of Afghanistan and Iraq.

Consider other examples along the line of moderate risk in Figure 1. Continued action against the bellicosi, whoever and wherever they are: high probability, low severity (as long as we stop the proliferation of nuclear, biological, and chemical weapons). Major international engagements, along the lines of the two Gulf Wars and which may, eventually, include limited nuclear exchanges: moderate probability, moderate severity. Full-scale nuclear war: low probability, high severity.

Before we go on, since we have referred to it several times now, let us take a few moments to consider NATO's war in Kosovo in some detail.

NATO's War in Kosovo

The NATO action in Kosovo in the 1990s was a low-severity conflict, but it represents the kind of conflict that can have a fairly high probability of occurring—because it seemed to be *popular*.

Few people at the time, other than some largely ignored critics and the leaders of Yugoslavian forces, noted the illegalities and inconsistencies of NATO's war in Kosovo. In retrospect, it is difficult to know whether the fear of a "spillover" into greater Europe or true concern for the displaced and downtrodden led to the vehement calls for military intervention. Regardless, if so many nations—including the United States—supported the effort, did that not legitimize it? Not on legal grounds, it would seem.

NATO's own charter, the North Atlantic Treaty, states that NATO will defer to the United Nations Security Council in disputes involving non-NATO nations. Article One states that the member nations agreed to settle international disputes by peaceful means, "and to refrain in their international relations from the threat or use of force in any manner inconsistent with the purposes of the United Nations."[11] Article Four requires the members to consult one another when one of the members' "territorial integrity, political independence or security . . . is threatened." Article Five enjoins the members to consider an attack against one to be an attack against all, and adds that, "Any such armed attack and all measures taken as a result thereof shall immediately be reported to the Security Council. Such measures shall be terminated when the Security Council has taken the measures necessary to restore and maintain international peace and security." And most importantly, Article Seven states that, "This Treaty does not affect, and shall not be interpreted as affecting in any way the rights and obligations under the Charter of the Parties which are members of the United Nations, or *the primary responsibility of the Security Council for the maintenance of international peace and security*" (emphasis added).

Yugoslavia, remember, still existed in 1999—several provinces had become independent, but neither Yugoslavia (now Serbia and Montenegro) nor its former provinces were members of NATO.[12] Almost all analysts overlooked the fact that the NATO action was not taken against a sovereign nation that had invaded another sovereign nation (in the way Iraq had invaded Kuwait), but was directed against one player in what amounted to a sovereign nation's civil—i.e., internal—war, presumably because Kosovo was a semi-autonomous province. These facts were also ignored by most pundits, politicians, and military leaders.

Later, US Congressman Ike Skelton wrote with apparent pride about NATO's "first military campaign," without seeming to bat an eye at the inconsistency of a defensive alliance taking offensive action to quell (but not settle) a neighbor's internal dispute.[13] The March 2004 announcement of NATO's expansion also included laudatory remarks on NATO's action in Kosovo and earlier actions in Bosnia.[14]

But to what ultimate end? As John Keegan wrote in 1994,

Prospects of peace-making may be illusory. The Balkan and Transcaucasian conflicts are ancient in origin and seem to have as their object that 'territorial displacement' familiar to anthropologists from their study of 'primitive' war. Such conflicts by their nature defy efforts at mediation from outside, since they are fed by passions and rancours that do not yield to rational measures of persuasion or control; they are apolitical, to a degree for which Clausewitz made little allowance.[15]

We may take Keegan's words cautiously—later, after all, he credited the victory entirely to airpower (see Chapters 2 and 10). But his earlier pronouncement was confirmed by the unfolding history: as of early 2018, the United Nations Interim Administration Mission in Kosovo still maintained a presence in the country.

In early 1999, when violence against the Serbians was being considered, few pointed out the hypocrisy in the US-led NATO action. We acted on behalf of, and supplied arms to, the Kosovo Liberation Army, a group previously branded by the US State Department as terrorists. This fact was conveniently overlooked, and not frequently reported. The objections of some of NATO's own member states (e.g., Greece) were ignored or downplayed, bringing into question who the members of the alliance really were. Perhaps all this was only an unfortunate example of "the enemy of my enemy is my friend."

In the end, the war crimes charges Yugoslavia brought before The Hague were dismissed, NATO settled into the peacemaking role, and Slobodan Milosevic went on trial for war crimes of his own. While some critics hyped that the Kosovo actions were curiously imitative of the movie *Wag the Dog*, the entire affair seemed more reminiscent of the "El Salvador—A War We Can Win" T-shirts that used to be sold in magazines like *Soldier of Fortune* in the early 1980s.

Some people may be prepared to defend the earlier Kosovo campaign on the grounds that the 2003 invasion of Iraq was the same: a coalition invading a sovereign state that had not (recently) taken any hostile action against its neighbors. While a case may be made, such an argument neglects two simple facts: first, the preponderance of authorizing statements from the UN Security Council that were notably absent from the Kosovo affair (attempts to secure UN authorization failed, and UN Resolution 1244, which among other things established the UN Mission in Kosovo, passed just as hostilities were winding down); and second, the numerous cease-fire violations by the Iraqi regime from 1991-2003 that made the Iraq situation more of one long, drawn-out war instead of two short wars.

A better defense of the Kosovo case may be on moral grounds: the rescue and return of the largely defenseless Albanian Muslims, rather than letting them perish or languish as refugees. That is a dangerous platform, however, since "ethnic cleansing"-type brutality in many other places (e.g., Rwanda, the Sudan, and elsewhere) has been far more severe and extensive, though largely ignored by the Euro-centric power system. And it is a false morality, since more Albanians were killed or displaced after the bombing than before; indeed, since the Albanians were not granted autonomy, NATO had "intervened in a civil war and defeated one side, but embraced the position of the party it had defeated on the issue over which the war had been fought."[16]

We leave the reader to ponder if the single UN Security Council resolution authorizing NATO intervention in Libya in 2010 truly provided sufficient justification for the US and Europe's defensive alliance to take offensive action against a sovereign, if despicable, government on the continent of Africa—even farther from the North Atlantic.

To return to the topic of the overall risk of war: if the risks are moderate, as shown in Figure 1, what individual risks do we try to mitigate? For example, do we ratify the Comprehensive Nuclear Test Ban Treaty, when the *risk* of nuclear war is probably no greater than that of a border war or trade dispute? Do we err on the side of increased probability, preparing to fight many small-scale actions, or do we err on the side of increased severity, attempting to forestall escalation of the nuclear conflict from the continued "cold" standoff to any "hot" war? Given our long engagement in an undeclared war against a barbarous enemy, it seems the decision has, for the time being, been made for us. As with any engagement, we must understand the situation and practice the art of command, willing to make decisions and predictions that may turn out to be wrong.

So, men will fight wars. That is, men will fight to assert some measure of dominance over other men, over other tribes or nations. Every battle, whether fought on land, or sea, or in the air, or (eventually) in space or the electronic realm, is an attempt to dominate the enemy, to force the enemy to submit to our will. This is true whether we want the enemy to behave in a way we determine, or we want to appropriate something from the enemy; and it is true whether we intend to dominate the enemy forever or only for a short time. It is not politically correct to couch war aims in such terms, but political correctness and truth do not always inhabit the same spheres.

4

WHAT WE KNOW, AND HOW WE KNOW IT

The subtitle of this book refers to what may be thought of as a "*military epistemology*." As Alexander Moseley states in *A Philosophy of War*, "Epistemology is the study of how we know things, what the limits to our knowledge may be, and how reliable is our knowledge."[1] Our study of what we think we know about war, and how we know it, is bound up in a term that military practitioners so often curse: theory.

What do we mean by "theory"? Once we stand on common ground, and establish some shared understanding of theory, we can go on to examine various military ideas to see whether they fit the bill. (Spoiler: Many of the ways we have organized thoughts about the nature of war and warfare—e.g., the levels, media, and principles of war—do not rise to the level of reliability expected of sound theories.) Those that fit the bill, especially if there are common threads between them, may be woven into doctrine, as "essentially sound doctrine can in fact be developed from a forward-looking theory."[2]

In the broadest possible terms, theory can be thought of as the mechanism by which we make sense of the world. More specific within the context of this work, theory is the attempt to explain the causation of past events or the condition of present circumstances; beyond that, theory is the foundation on which we build our predictions of the probable outcomes of events.[3] This epistemology—Theory of Knowledge—is detailed in US pragmatic philosopher C.I. Lewis's *Mind and the World Order*, in which Lewis wrote, ". . . it is impossible to escape the fact that knowledge has, in some fashion and to some degree, the significance of *prediction*."[4] In applying theory to warfare, then, we explain the results of past conflicts in ways that, hopefully, predict the outcomes of those yet to be fought.

But theory is not only *useful* for explaining the past and predicting the future: it is *necessary* to it. Theory is not a substitute for genius and judgment as stated in the old AFM 1-1: it is a necessary element of genius and judgment. Only by applying a theory can we make the past more than a collection of facts, can we answer "Why did this happen?" instead of "What happened?" Indeed, developing theories (or concepts, to use C.I. Lewis's preferred term) is necessary to having any knowledge at all:

> In experience, mind is confronted with the chaos of the given. In the interest of adaptation and control, it seeks to discover within or impose upon this chaos some kind of stable order, through which distinguishable items may become the signs of future possibilities. Those patterns of distinction and relationship which we thus seek to establish are our concepts. These must be determined in advance of the particular experience to which they apply in order that what is given may have meaning. Until the criteria of our interpretation have been fixed, no experience could be the sign of anything or even answer any question. Concepts thus represent what mind brings to experience.[5]

"The given" are the facts with which we come in contact: they are either historical facts, or facts (such as we know them) about our current situation. And in the main "the given" is a chaotic mess.

Think of all the inputs that bombard us minute-by-minute, day-by-day, even while you are reading this text: sights, sounds, and other sensory signals, many of which our brains simply ignore because they are not vital to what we are doing at the time. In many respects, this filtering process is one of the most important differences between humans and animals.[6] Some of these myriad impulses get through the filters of our senses and into deeper reaches of our minds, where we consider them in greater detail. We treat facts of history the same way as these facts of immediate life, ignoring those we do not find immediately useful but storing some away for deeper thought or reference at a future date. And in the process, in our minds we develop and apply patterns and concepts with which to make sense of it all and provide a basis for action based on future possibilities. Without interpretation using theory, then, facts have no predictive power.

The fact, taken alone, that Iraq had in 1990 the world's fourth largest army gave no insight into the eventual overwhelming victory of the coalition forces in the first Gulf War. The fact that several men of Arabic descent engaged in suspicious activities involving flight training did not by itself indicate that 11 September 2001 would be any different from 10 September. The fact that the United Nations Security Council passed over a dozen resolutions calling on Iraq to disarm, beginning with an April 1991 resolution requiring the destruction of Iraq's chemical, biological, nuclear, and ballistic missile programs—including subsequent resolutions that noted Iraq's "serious violation" (August 1991), "clear violation" (June 1996), "clear and flagrant violation" (June 1997), "flagrant violation" (October 1997), "totally unacceptable contravention" (September 1998), "flagrant violation" again (November 1998), and finally "material breach" (November 2002)—gave no insight into how difficult it would be to find clear and indisputable evidence that their previous weapons programs constituted a more immediate danger.[7] The fact—again, taken alone—that *you are reading this text* gives neither of us any insight into what you will do when you are finished. Only by interpretation can we satisfactorily take facts into account; only applying theory allows us to predict the future with any probability. "The theory of knowledge teaches us that a statement, if it conveys knowledge, predicts future outcome, with risk of being wrong, and that it fits without failure observations of the past."[8]

It may come as a surprise that theories of war could follow and incorporate elements of general theories of how the world works. In *The Links Between Science, Philosophy and Military Theory*, Robert P. Pellegrini wrote,

> General histories of science and of western civilization usually show the connection between science and philosophy, but never their impact on military theory. Military histories usually discuss in general terms the philosophical influences which affected military theorists but seldom discuss the scientific genesis of these influences.[9]

But evaluating how the world works is not the exclusive domain of the scientist. "Essentially, the role of the scientist in this process has been to describe the nature of the universe. The philosopher then described man's place in this universe. Finally, the military theorist described how man best conducts war in this universe."[10]

While this separation may hold true for the deepest investigations into the world, it does not hold true at the "everyman" level. Everyone, carpenter to corporate executive, steel worker to secretary, bricklayer to brain surgeon, must make basic evaluations and predictions to survive in the world. We do not match the scientist in her precision or the philosopher in his depth of thought, but we do not need to in order to make necessary sense of the world in which we live our lives. In a sense, everyone is their own personal scientist and philosopher—"everyone *can* be his own philosopher, because in philosophy we investigate what we already know"[11]—in that same sense, every military member is their own personal military scientist and philosopher of war. And we philosophize on the basis of theory.

Theory allows us to categorize facts and experience in such a way that we may make sense of them for future use. From the conditions we face and the experiences we recall, we formulate ideas about what may happen; then we evaluate our subsequent experiences and learn whether our ideas were correct. Some of our predictions may be vague, others very precise; for example, on the personal level we may state that two years from now we will be out of debt, or we may state that on a specific future date we will make a payment of a specific amount of money that eradicates our debt. We choose which level of precision to use based on our confidence in the prediction, but always we run the risk of being wrong. Using the debt example, we may be wrong because we figured the interest incorrectly, or because we failed to account for hidden charges, or because some future crisis causes us to forego debt payments while we correct the immediate problem. If we err, it is not a tragedy, even if it has unfortunate consequences. Indeed, it is possible to use failures as well as successes to refine our thinking with respect to other predictions we will make. Each successive prediction still has the possibility of error, which is part of what differentiates knowledge from faith: with knowledge, "In general there must be the possibility of arguing from past to future; not with certainty, but with probability."[12]

It is one thing to make statements about the future that have the possibility of being wrong; it is another to make such statements about the past. Insofar as the past is verifiable, our statements about it should be consistently correct. The possibility of error in our statements about the past arises because we missed some salient fact or our analysis of the available facts is wrong; in contrast, if our statements convey real knowledge of the past then we must have weighed the evidence correctly and our thinking must be sound.

To reiterate: When we encounter an experience or a collection of facts, we develop a theory or concept to explain it so we may understand it. That theory then becomes the basis for predictions of further experience, although the act of predicting may include an emotional component as well.[13] This process occurs so automatically—and has from the time we were infants—that we are largely unaware of it, but it explains why "we do not see patches of color, but trees and houses; we hear, not indescribable sound, but voices and violins."[14] If we are to be able to learn, we must be able to categorize experience in this way, and that ability then allows us to predict how upcoming experiences are likely to develop. This, for example, is why "principles of war" were developed in the first place.

If we came to each new experience with no ability to interpret it by means of theory, we would be unable to make sense of the new experience. If we had no theory to begin with, it would be as if we had no memory; the past would be a mishmash of images and emotions that would add nothing to the current experience to enable us to predict any future outcomes. Alternately, if we had no ability to formulate some *new* theory to explain our experience it would be as if we had no mind: experiences would happen to us and bombard us with stimuli but we could make nothing out of them.

In terms of military experience, Prussian theorist Carl von Clausewitz warned that "as soon as preparations for a war begin, the world of reality takes over from the world of abstract thought,"[15] but we still need theory in order to interpret the world of reality. The explanatory theory gives us insight into the past, which aids our present understanding and in turn gives us a basis upon which to predict the future: as C.I. Lewis points out,

> . . . there is no knowledge of external reality without the anticipation of future experience. Even that knowledge implied by naming, or the apprehension of anything presented, is implicitly predictive, because what the concept denotes has always some temporal spread and must be identified by some orderly sequence in experience.[16]

The explanatory theory thus lays the conceptual groundwork for the predictive theory.

Though important, explanatory theories do not generate new knowledge; they only arrange existing knowledge. The predictive theory is necessary to generating new knowledge, to "[shed] light on problems and thus . . . provide guidance for those who have the responsibility for solving them."[17] In the same vein, Swiss-born French General Antoine Henri Jomini wrote,

> It is true that theories cannot teach men with mathematical precision what they should do in every possible case; but it is also certain that they will always point out the errors which should be avoided; and this is a highly-important consideration, for these rules thus become, in the hands of skillful generals commanding brave troops, means of almost certain success.[18]

Note Jomini's promise of "almost certain" victory as contrasted with the Chinese military philosopher Sun Tzu's notation that one can (but not necessarily will) win by making many calculations. But the point here is that we bring to each new experience a theory of how we expect it to turn out, without which we would almost certainly be powerless to act, and would instead be acted upon: "What lies on the surface can be taken as ultimate only so long as there is no problem to be solved, or else no solution to be hoped for. *Without analysis, there can be no advance of understanding.*"[19]

5

THEORY IS THE KEY TO UNDERSTANDING

We develop initial theories in rudimentary form, then refine them by comparison with actual experience. That may be direct experience, or vicarious—thus the possibility of learning from others' mistakes as well as our own. But as facts alone have no predictive power, so examples of experience by themselves do not comprise a theory: experiences, like facts, must be interpreted.[1] Examples of battles won or lost are useless without theories to explain the causes, and every battle plan or Air Tasking Order, though directive in nature, is implicitly a predictive theory of how events are expected to pan out.

Indeed, every decision we make is based on a theory that we have either adopted or developed. For example, you may decide to drive to a concert or sporting event based on your friend's prediction that traffic will be light; or you may develop your own assessment, predict a different outcome, and decide to take a taxi, an Uber, or some form of mass transit.

We usually do not think of such commonplace choices as being based on theory, because we have assigned "theory" a place beyond the ordinary, a higher plane where only scientists and philosophers operate. One way to correct this notion was mentioned in the last chapter: specifically, for every man and woman to be their own philosopher. A better way may be to demystify "theory" itself.

If we change our viewpoint or the distance from which we observe something, we may see more or less of the whole or see new aspects of it. Seeing it anew, in different light or up close or in relation to something else, we may derive different knowledge of it. The proverbial elephant seen as a whole rather than felt by several blind men comes to mind, and is a useful metaphor of different methods of perception. Another example, though trivial, may also help illustrate the concept.

Imagine that you and your friend are standing a football field's length away from each other. Your friend holds a slim object between her thumb and forefinger such that it looks like a short vertical line:

|

If your eyes were tuned to more wavelengths of light than they already are (for those with interest in such things, the human eye *is* a multispectral sensor, else you could not distinguish colors), you might be able to discern what color it was and perhaps then what it was made of, but at this point you have no other information than its size (very small) and shape (a slender rectangle). Without additional information you must rely on theory—based on experience—to form any idea of what she is holding. Your ability to infer a number of possible candidate objects will be affected by your visual acuity, the lighting conditions, and many other factors, as well as your experience with small, thin objects. If she turns the object in her fingers such that you now see it from another angle and it looks like a circle:

O

the additional datum that it is round in one plane and flat in another makes it easier for you to predict what she holds. At this point you may pronounce that she is holding a coin, but you will require more information and closer inspection to specify what type of coin she is holding. And your knowledge that she is holding a coin may turn out to be erroneous: she may be holding a washer, a game token, or a slug.

According to Theory of Knowledge, even naming the item as a "coin" rather than a "slender metallic-looking disk" forms an inherent *prediction* of its future usefulness based on the different uses of coins as opposed to washers, game tokens, or slugs.[2] We do not need to go through any great mental gymnastics to make such predictions—i.e., to develop such theories. Each of us does it a thousand or more times a day, without conscious attention, and would find life nearly impossible if we did not have the ability or the inclination to do so or if we had to make the predictions consciously. Theory, then, is the natural basis for understanding, for negotiating life at any level higher than pure instinct, and for explaining the past and predicting the future.

Most of our "theorizing"—i.e., developing theories to explain life's experiences—is unconscious (or perhaps pre-conscious) for the more common, less significant bulk of life, in that we do it without the intervention and benefit of conscious thought. Still, it is a cognitive function by which we develop predictions and test experiences without stopping to think about how we do it, or why. We do not bother to compute the probability that the earth's rotation will change such that the sun will not appear to rise in the morning; we merely predict that the sun will rise and get on about our business. The process of theorizing is innate to thought, and to life.

In fact, we encounter and use theory in every facet of our lives. Theorizing is so natural to our knowledge of the world that we rarely recognize it; even so, developing concepts in this way to explain and predict experience drives all our knowledge.[3] In the same way, theory provides an indispensable framework to understanding the reality of war. As US Navy Rear Admiral Henry E. Eccles wrote,

> The difference between the theory and the reality of war, just as the difference between the theory and the reality of anything, is like that between a skeleton and the living man with all his flesh and blood. Compared with a man, the skeleton appears strange; but we could not understand how his limbs cohere and function if we did not penetrate his skin to the skeleton and its joints.[4]

The natural and unconscious process of developing everyday theories—unconscious in that we do not think, "I am now developing a theory to explain this"—may be represented by the chain reaction

experience → concepts → theories → applications

where applications then affect subsequent experiences. That is, our actions based on our theories will lead us into new and different experiences, which thereby lead us to develop different concepts, and so on. Similarly, the process of

conflict → concepts → theories → applications

has worked for thousands of years of explaining and understanding battles, wars, and warfare. The *first* battles, maybe even the first wars, may have been fought without the benefit of theory, but only the very first.

Men experienced a battle or war ("noise and confusion, and garments rolled in blood"); the survivors pondered what happened, and sometimes how they survived: "oftentimes it turns out that even the participants are not sure what happened."[5] Those survivors who continued military or government careers ferreted out lessons learned or even "principles" to explain what they experienced; those lessons or principles were applied, to whatever degree of success, in the next encounter—sometimes after years in which the lessons became, truly, *lessons* taught to students of war. If the lesson resulted in victory, or at least survival, in that subsequent moment of trial, it would continue to be taught and reinforced. If it continued to prove successful more often than not, it would be virtually assured a place in the military canon.

Because the process by which lessons such as these evolve is first unconscious—or, if you will, automatic—and second (in many cases) quite lengthy, their status as *theories* may go unrecognized. Each lesson becomes accepted as truth, immutable and unquestioned law, rather than a theory that may be flawed. Much of this book is intended to put some of these "truths" to the test.

When we test our theories against experience, we sometimes find the need to change them. In our thought experiment above, our knowledge that our friend is holding a coin will have to change if she hands us a washer. In the same way, a 1995 article on information warfare stated, "If the facts get in the way of a theory, then the theory should be discarded, not the facts of history. . . . One does not base grand theories on false facts; nor does one prepare for the future by distorting the past."[6] If the theory fails outright, and some unfortunately do, it must be discarded in favor of a new theory. The history of science is full of examples of this process, such as the abandonment of Ptolemaic cosmology in favor of Copernican. Again C.I. Lewis uses the word "concept" when he states, "When particular concepts fail, we merely abandon them—through analysis or organization or abstraction, and so on—in favor of corrected ones, which take cognizance of, and include the ground of, our previous failure."[7] The difference, of course, is that everyday theories that fail usually do not leave us responsible for death and destruction: so military theorizing must be that much more careful.

Sometimes our theories are incomplete, rather than fatally flawed, and fail only in a few cases. These theories can be refined rather than abandoned, as Albert Einstein refined and expanded his original Special Theory into his General Theory of Relativity. Comparing theory to experience, and evaluating and possibly altering the theory, is a continuous process, because "what experience establishes, it may destroy; its evidence is never complete."[8] In other words, it is always possible that the next experience may prove the theory wrong.

When it comes to military applications, since we contend not with the relatively static natural world but with forces that actively oppose us, we must not lose sight of the role of theory as a guide to understanding. As US Army Lieutenant Colonel Tom Kratman put it, "It may be that no theory about war is worth anything except as a study guide, useful to form the minds of the combatants and then discarded as their aptitudes, both learned and innate, take over."[9]

6

THEORY IS LIKE A GOOD ATTITUDE

Renowned motivational speaker Zig Ziglar used to say that a good attitude would not help anyone do anything; however, it *would* help anyone do anything *better* than a bad attitude would. Theory is much the same. A good theory does not explain everything. But a good theory will explain things better than a bad theory.

It behooves us, then, to make our theories as good as possible—especially when we are considering theories about something as important, as monumental, as consequential, as war.

Our knowledge of war is fragmented. Because of the exponential growth of general knowledge as well as martial knowledge, we tend to focus our studies on precise subjects of interest. We develop specialists in the art of war—tacticians and logisticians, operators and support personnel—and our approach *in toto* becomes a combination of separate specialties. Our specialties, while necessary, separate us one from another. Though we strive to develop "general" officers, few really are.

What we lack is common ground, a common thread between specialties, a theory compatible with our different fields that takes elements of existing theories and ties them together to produce a more complete picture. The epitome of this type of composite theory is the long-sought unified theory of physics that will tie together gravitation and electromagnetism and explain the formation and behavior of everything from quarks to quasars.

Is a unified theory of war possible?

War, we recognize, is not physics. That may seem apparent, if not unnecessary to state; but if physics may be thought of as the collected observations about the mechanics of the world, together with mathematical formulas that describe those mechanics, then we might at least admit the possibility of a branch of military science—"martialists," if you will—that would study war the way physicists study the physical world. Their study would not involve direct experimentation, of course, because it would be impossible and most undesirable to actually fight the same battles over and over, changing certain variables here and there to see how the outcomes shift.[1] Computers may make it possible to *virtually* re-fight the same battles many times, changing variables here and there, but the conclusions would be only as good as the models and simulations used. Still, these martialists could collect observations about the mechanics of war, and might even develop mathematical or other formulas that describe those mechanics.

"Other" formulas or descriptors—that is, non-mathematical descriptors—would be especially important in such an esoteric study of war and warfare. War, in which human forces clash, does not lend itself to mathematical certainty or even precision in the way science does. That also may seem self-evident, but that has not stopped some people from attempting it. Jomini's insistence on one true principle of war could, for instance, be stated mathematically as

$$P_V \ \alpha \ C_{(t,p)}$$

or, the probability of victory is directly proportional to the concentration of forces at the decisive time and place—making Jomini perhaps the quintessential martialist of his day.

A present-day martialist trying to apply mathematical precision to warfare might expand this equation in an attempt to derive something analogous to a gravitational constant of war—a value that could be applied to any situation to determine the probability of victory. But war is not amenable to such precise notation. War has too many variables and their interactions are too complex. Even so, a unified military theory that could tie together other theories is a potent idea. It would be remarkable if the range of military theory, experience, and wisdom could be distilled into a single package, even if that package were cumbersome.

Though it is implausible, the mental exercise of searching for a unified theory of war may benefit us regardless of the outcome. Combining existing theories into new patterns should improve our ability to plan and execute, as well as predict the outcome of, campaigns and battles. For those of us whose natural aptitude for war will never measure up to history's great commanders, this exercise, even if it ultimately fails, may prove worthwhile. We hope it may give us at least a glimmer of that quality known as *coup d'oeil.*

This "stroke of the eye" is the ability of a commander to see through the trivial to the critical elements of a battle. Miyamoto Musashi wrote, "The principle of strategy is having one thing, to know ten thousand things,"[2] but coup d'oeil may be better thought of as, "Having ten thousand things, know one thing." Both statements are correct, and both are necessary: to be able to glean from a single salient factor a myriad of possibilities, and to be able to sift through the "chaos of the given" to find the most important nugget of truth.

Clausewitz appreciated the advantage *coup d'oeil* conferred on the warrior. He elaborated on the topic:

> If the mind is to emerge unscathed from this relentless struggle with the unforeseen, two qualities are indispensable: *first, an intellect that, even in the darkest hour, retains some glimmerings of the inner light which leads to truth; and second, the courage to follow this faint light wherever it may lead.* The first of these qualities is described by the French term, *coup d'oeil*; the second is *determination.*[3]

This ability to cut through the mass of incoming impulses is necessary to more than warfare: it is necessary to thought. What military thinkers know as *coup d'oeil* was discussed in more general terms by C.I. Lewis: "Confronting any given experience, the first act of intelligent cognition is to discard all but a few items of what is presented as excess mental baggage irrelevant from the point of view of our predictive purpose."[4]

Jomini, agreeing with Clausewitz, also noted the importance of *coup d'oeil.*

> In this important crisis of battles, theory becomes an uncertain guide; for it is then unequal to the emergency, and can never compare in value with a natural talent for war, nor be a sufficient substitute for that intuitive *coup d'oeil* imparted by experience in battles to a general of tried bravery and coolness.[5]

Had Jomini recognized the necessity of theory, he might not have disparaged it so, but he was right that theory is a guide and never a perfect one. In his *Note upon The Means of Acquiring a Good Strategic* Coup-D'oeil, Jomini called *coup d'oeil* "the most valuable characteristic of a good general, without which he can never put in practice the finest theories in the world."[6]

Coup d'oeil, then, even if rudimentary, is necessary to translating theory into reality. We plan and train based on the known, but war is full of unknowns that make themselves evident at the most inopportune times. A firm grasp on theory allows us to evaluate and act on the infinite inputs war presents us, to "discard [the] excess mental baggage." With that in mind, we must have a firm grasp on how theory relates to experience. We can then examine different theories, test them against reality as we know it, and possibly combine them into new descriptions of warfare.

Some may argue that this exercise is impractical, or even too "intellectual." Perhaps so. Our intent remains otherwise, however, since "the ruling interest in knowledge is the practical interest of action."[7] We study and theorize in order to act, and in order that our actions will be effective.

7

PUSHBACK

Throughout history, theories and concepts of warfighting have adapted to new circumstances, enemies, and weapons. This process continues today, and the evolution of martial theory may possibly lead to some unified theory of war and warfare. But because war is such an unstable enterprise, so susceptible to chance and personality, it is at least as likely that new developments may make the search for any unified theory futile. However, because such a search involves a critical analysis of existing theories, it may still yield valuable results: "theory can never lead to complete understanding, which is an impossibility, but it can strengthen and refine judgment."[1]

It may *seem* sometimes that we have failed to learn the lessons of history, but some lessons are harder to learn than others, in more ways than one. Many writers have pointed out that we ignore the lessons of history to our peril, but we must be careful as we proceed: "History is by no means obsolete; it is still capable of offering us important and highly useful perspectives, but we must be everlastingly cautious in using history, for it is seductively easy to misinterpret and misuse."[2] Therefore, we must fight the temptation to glorify the past in our attempts to learn from it. "To forsake the knowledge of past wars would be foolish; to try to fight the next war as the last war was fought is also foolish."[3] US Air Force Colonel John Warden, architect of the Desert Storm air war, put it more succinctly: "the old ways may be a recipe for disaster."[4]

To avoid falling into this trap—and to be consistent with Theory of Knowledge—theories that are outdated or yield inconsistent results need to be reformulated. Current, prevalent and popular theories of war may fall into this category; we will analyze some of them in succeeding chapters, but before we go any further we should consider dissenting views in the form of some prominent thinkers' objections to this emphasis on theory, especially as it pertains to military matters and martial action.

As discussed so far, every action we take is based on some prediction we made about the outcomes of that action; for our purposes in this book, we call that process "theorizing." Note, though, that prediction does not mean absolute foreknowledge: the play of probabilities and the intrusion of unplanned events mean knowledge may be erroneous and predictions incorrect.

Jomini, though he did not have this epistemology, argued against one of its basic tenets. He wrote, ". . . a single exception cannot disprove a rule based upon the experience of ages and upon natural principles."[5] That is, according to Jomini a rule may continue to be accepted as a rule *even though it falls short in particular cases*. Theory of Knowledge teaches otherwise: the exception cannot really "prove" the rule, and must be accounted for in the theoretical construct. Jomini failed to consider that exceptions require explanation, and explanations are innately modifications to the rule (i.e., the theory) itself.

In another passage Jomini asked,

> When the application of a rule and the consequent maneuver have procured victory a hundred times for skillful generals, and always have in their favor the great probability of leading to success, shall their occasional failure be a sufficient reason for entirely denying their value and for distrusting the effect of the study of the art? Shall a theory be pronounced absurd because it has only three-fourths of the whole number of chances of success in its favor?[6]

This question is very similar to Liddell Hart's contention that "if a specific effect is seen to follow a specific cause in a score or more cases, in different epochs and diverse conditions, there is ground for regarding this cause as an integral part of any theory of war" regardless of whether it fails in the twenty-first—or for Jomini, the hundred and first—example.[7] Contrary to Jomini's insistence, explaining "occasional failures" by modifying a theory is not equivalent to "entirely denying" the value of the theory; indeed, it would seem to enhance its value and strengthen "the study of the art." No, pronouncing a theory "absurd" is not in order when its failures reduce its probability of success, but neither are codifying and adhering to it as if the theory guaranteed success. It seems more absurd to observe the failure of a theory and then pursue the same course of action on the grounds that it had only failed once.

For an everyday example, we might think of the absurdity of seeing a light bulb blow out but continuing to turn the lamp on and off on the grounds that it worked every time *except* the last time. From a military standpoint, we might think of the futile Confederate charges at Gettysburg, which, having failed—but being similar to charges that had been successful before—were repeated until the Rebels were hopelessly worn down.

When applying his maxims, some of Jomini's successors apparently missed his reference to the probability, rather than the certainty, of success. They may be forgiven since Jomini himself wrote of "almost certain success" if his maxims were followed. Liddell Hart did not fall into that same trap, but he still succumbed to the temptation to allow for exceptions rather than revise the theory. All of which supports his own statement that "the only thing harder than getting a new idea into the military mind is to get an old one out."[8]

Despite pushback from military thinkers and practitioners who cling to ideas and methods which have worked in many (but not all) cases, according to Theory of Knowledge those ideas that fail to produce victory or survival should be modified—and even, at times, abandoned. Unfortunately, as we will see, military theories, unlike old soldiers, rarely "fade away."

8

IS WAR UNDERSTANDABLE?

As noted in Chapter 2, "war" is not defined in the DoD's official dictionary; neither are "battle," "combat," or even "conflict." In contrast, the previous version at least defined "conflict," as an "armed struggle or clash between organized groups within a nation or between nations in order to achieve limited political or military objectives."[1] This definition resembled to a curious degree what many people may think of as "war." It is especially curious that the definition referred to an *armed* struggle, as if nonviolent, unarmed conflict is not possible, although it makes sense that the Defense Department would concern itself first and foremost with armed struggles. In contrast, Alexander Moseley defined war as "a state of organized open-ended collective conflict," and elaborated: "War is organized, unlike, for example, a street brawl; it is open-ended, unlike a boxing match; it is collective, unlike a personal feud; and it involves conflict that implies nonviolent as well as violent hostilities."[2]

Even though all war is conflict, all conflict is not necessarily war. Once again, let us start with the most basic ideas. *Violent* conflict (to be specific) may be described as an armed force encountering and engaging an opposing force as they try to achieve an objective. In this academic description, the nature of the objective is not important: it may be physical or tangible (e.g., land or other resources), economic, ideological, or even illusory. Applying this definition, we may conclude that both an opposing force and an objective are necessary for war, or at least for battle. Without an opposing force there will be no conflict and certainly no war, because we could then pursue (if not achieve) the objective with little or no effort. Without an objective there will also be no conflict or war, because we would have no reason to fight.

Wars have been fought over trivial objectives, poorly defined objectives, or objectives even the combatants would find difficult to identify, but that does not negate the fact that there was some underlying objective and at least one overarching aim: defeating the enemy. (In another vein, if an objective is a necessary condition for war, that may invalidate "objective" as a "principle of war"—a topic we will examine in detail in Chapter 19.) Identifying the objective is critical to success, however, as pointed out by Clausewitz:

> No one starts a war—or rather, no one in his sense ought to do so—without first being clear in his mind what he intends to achieve by that war and how he intends to conduct it. The former is its political purpose; the latter its operational objective. This is the governing principle which will set its course, prescribe the scale of means and effort which is required, and make its influence felt throughout down to the smallest operational detail.[3]

Most readers are no doubt aware of Clausewitz's more famous line regarding war as it relates to politics, although translation issues mean this itself is something we may only think we know. While it is popularly translated as some variation of "an extension of politics," a more complete rendering is "war is the continuation of political discourse with the addition of other means."[4] Martin van Creveld points out that this view of war "is in some ways a modern invention,"[5] and one way to understand it is to realize that a war's objectives (there may be more than one) are inherently political in the sense that they are—or should be—in the interest of the *polis*, i.e., the people and institutions comprising the nation.[6] We would do well to be cautious in how we interpret and apply things that derive from translations; they may not capture all the nuances of the original.

The broad objectives listed earlier—physical, economic, ideological—take a variety of forms and combine in numerous interesting ways, but can be summed up in a simple phrase: We fight over the things we value. That is, we only fight for those things we consider valuable. Freedom, land, honor, love, money, power, pride, etc.: if we value it enough we will fight to obtain it or keep it. As van Creveld notes, "Since he who fights puts everything at risk, whatever he fights for must be deemed more precious than his own blood."[7] We may address all objectives under this rubric of value, including (or perhaps especially) economic objectives, although the US quasi-intelligentsia—media, celebrities, pundits—usually imply (if they do not state outright) that the nation's economic goals are somehow not legitimate. (How this anti-capitalistic viewpoint developed is beyond the scope of this work; it may represent a general anti-establishment, anti-power bias grown from the idea that only "revolutionary" goals are legitimate. It would be interesting to investigate this along sociological and/or socio-historical lines.)

To reiterate, then, all war is conflict in pursuit of some set of shared objectives; but all conflict is not war. The difference is violence, and it is no mistake that the underlying theme of this discussion has been the application or threat of violence: "Force is the basis of any military activity, whether in a theatre of operations or in a skirmish between two soldiers."[8] War, as we commonly think of it, is violent activity. The application of violence—and especially deadly violence—separates "war" from "warfare," i.e., the state of war from the practice of war.[9]

We may well ask whether there are *degrees* to which violence is necessary to war. With respect to terrorism and General Smith's "war amongst the people," for instance, former US Senator Gary Hart wrote, "New forms of violence resemble war, but by historic standards they are not."[10] Certainly a "battle" in the classic sense will be a violent affair. If violence is not necessary to war we might question whether a war could perhaps be fought—if "fought" is the right word in this case—without engaging in battle. We may, for example, think of violence as a "violation"—a coercive intrusion, but not always bloody. This idea fits especially well with the idea of "netwar," which we will discuss in more detail in Chapter 15. An electronic attack against industrial computer systems would not necessarily be violent in the common understanding, even though its results—intended or not—may be devastating.

Under some circumstances the threat of violence may be enough to secure victory. As Sun Tzu wrote, ". . . to win one hundred victories in one hundred battles is not the acme of skill. To subdue the enemy without fighting is the acme of skill."[11] In those rarest of historical cases in which one side capitulated without a violent struggle, the other side was always ready to engage.

The possibility of deadly violence separates war *per se* from the various other contests that have "war" appended to them. For example, countries and companies may contend with one another in "trade" or "economic" "wars," and may even employ some means of interdiction to prevent each other from achieving certain goals, but usually those contests do not involve violence or the threat of its use. The participants in such non-wars may attempt coercion, but not outright violence; for their actions to constitute or even approach "war" in any real sense, violence must be implied if not applied. In contrast, we may think of the drug "war"—in which selective violence is applied and the threat of further violence looms—and the Terror War, in which violence is met with violence and only the disingenuous believe complete nonviolence will prevail.

You may dispute (or even dislike) this reduction of war to very basic constituents. Admittedly, the topic leaves room for a wide range of alternate viewpoints: we may disagree over the definitions of constituent parts; we may replace some parts with others more to our liking; we may add parts; we may order or rank the parts according to one hierarchy or another; and so forth. The idea is not to declare categorically and forevermore what the parts of war are, but to describe some basic constituents with the aim of examining the nature of war and warfare as a whole. We start by taking a limited view, the way we would conduct a systematic telescopic study of the heavens by taking a section of sky at a time.

To recap, then, for the purpose of this discussion the basic constituents of war are: a force or instrument of power capable of violent action; an objective; and an opposing force. Without the aspect of violence these same constituents play into any struggle to obtain any objective, be it an attempt to climb a mountain—the power of strength and determination, the objective of the summit, and the opposing force combining nature, the elements, time and gravity—or the attempt to win any competition from a footrace to a board game to a sporting event. War, with the risk and danger inherent in the violence it involves, can be thought of as conflict on a grand scale, the highest stakes gamble that exists.

In the conduct of war, we must consider each constituent very carefully. We must evaluate whether our force is equal to the task and whether the objective is worth the risk and the price. Sun Tzu was quite adamant about the importance of calculations such as these: "With many calculations, one can win; with few one cannot. How much less chance of victory has one who makes none at all!"[12] The most important evaluations, however, must be our estimates of the opposing force; specifically, what the enemy is capable of doing and likely to do in any given situation.

We may think of the planning for war, then, as a series of if-then propositions in which we postulate how our enemy will react. The simplest such proposition will take the form of, "If we do this, then . . .", but by itself that form is hardly helpful. Will the enemy rise up to thwart us? Will their forces be confused and unable to respond effectively? Will other forces intervene? This last was the question Iraqi President Saddam Hussein failed to ask—or answered incorrectly—before invading Kuwait in 1990. Will they live up to their pledges, and enforce the resolutions they've enacted? This was the question Hussein failed to answer before the second Gulf War drove him from power in 2003. Will the enemy remain alert and responsive while distracted by other conflicts and economic troubles at home? This was one of many questions Osama bin Laden failed to answer correctly before Seal Team Six descended on his compound in 2011. We may create never-ending lists of similar questions.

Asking such questions in the first place is one thing, but in order to begin answering them we must add more and more detail: levels and levels of complexity that build up like layers of oil paint on a canvas. For example, we must consider what we want to do in light of what we believe the enemy is likely to find out, producing an advancing series of propositions that take the form of

> If we do this, and the enemy knows, then . . .
> If we do this, and the enemy doesn't know, then . . .
> If we fail to do this, and the enemy knows, then . . .
> If we fail to do this, and the enemy doesn't know, then . . .

Unfortunately, we find that the questions become unwieldy as we add more and more conditions. We want to make the questions meaningful and to include all the possibilities we think are important, but in the process we could build such a long series of questions—nested together like loops in a computer program or matrixed together like equations to be solved simultaneously—that attempting to apply it to a real situation would be madness.

Certainly questions like those above apply to war in its totality as well as to warfare, i.e., the conduct of operations, but to have any hope of success in a crisis we must be able to strip them bare of nonessential factors and answer them quickly and efficiently. In the heat of battle, during the clash with our immediate foes, we have no time for excessive meditations, and so we turn to our already-written doctrine to guide us. Doctrine—"fundamental principles that guide the employment of . . . military forces"[13]—may be thought of as a compendium of the techniques and practices that have worked in past encounters. Doctrine serves an indispensable purpose: to provide easily accessible guidance when the enemy are at the gates and the response is either to fight now or die.

But when the hordes clamor outside the walls, it is too late to try to understand war. If we are to survive the crisis, our best approach is to be prepared for it before it begins. As the Roman historian Vegetius wrote, "He who desires peace, let him prepare for war."[14] Preparing for war takes time and treasure and hard work, and unless we prepare well and carefully we run the risk of squandering them instead of investing them wisely. In times of relative peace, when we are not hard-pressed under the immediate rush of an approaching enemy, we can consider a multitude of options and alternatives. But to gain understanding we must know in advance what questions to ask. Whether they take the form of the simplified questions above or more complex models of international conflict, we must know which questions are most meaningful, which most efficient at revealing situations and most effective at generating alternatives. And despite the low regard in which it is held, *theory* helps us generate those questions and evaluate the answers.

9

HAS ANYONE UNDERSTOOD WAR?

Despite their sometime reluctance to admit it, warriors and other commentators have theorized about warfare ever since war began. In very basic terms, they developed explanations for victories and routs, and explored the characteristics of warfighting in order to predict success in battle. Although many of the more famous Western commentators would have been unfamiliar with the reference, their efforts aimed at fulfilling Sun Tzu's admonition from *The Art of War*, "Know your enemy and know yourself; in a hundred battles you will never be in peril."[1] As for the importance of their efforts, US Air Force Colonel Edward C. Mann III wrote,

> The duration and casualties of World War II could have been reduced if concepts and technologies had been better matched and/or the concepts themselves more fully developed. . . . But, as is usually the case, concepts and theories—not technologies and tactics . . . or brilliant individuals . . . were principally responsible for the Allied victory in World War II.[2]

If current military personnel find theory distasteful, they should not feel alone. Even some famous military theorists wrote disdainfully of theory, and presumably would not appreciate being considered theorists at all. But a great many military thinkers throughout the ages—including some of the most brilliant commanders—have seen fit to explore what factors improve the chances for success in warfighting. It behooves current military practitioners, policy makers, and those with a stake in the outcome to familiarize ourselves with at least some of their thoughts.

Sun Tzu (or Master Sun, or Sun Zi),[3] for instance, is generally recognized as one of the most ancient military theorists, having written his *Art of War* over 2000 years ago. The author's true identity is a mystery, but he was presumably a successful general during either the "Spring and Autumn" period or the "Warring States" period of Chinese history.[4] He carefully categorized various aspects of warfighting—e.g., terrain, weather, espionage—and included in his text observations on their many variables; suggestions for how the diverse factors influence formations and tactics; and basics with which a prudent commander may use them to his advantage. Many assume that the compilation that bears his name was not wholly original to him; regardless, its influence in challenging the "Western" way of warfare has been felt keenly, especially since China's Communist revolution.[5]

Carl von Clausewitz's name is well known in the military, even if his theories are not. Simplified and often misunderstood versions of Clausewitz's ideas—especially as related to politics, fog, and friction—are as prevalent as clearly articulated versions. That is because his *On War* is one of the most difficult monographs on military theory. Its difficulty is not simply a matter of translation; rather, it does not over-simplify the nature of war but presents it as a complex whole, a clash between living armies.[6] As Martin van Creveld put it, *On War* "is mainly deductive in character: starting from first principles, the nature of war and the goal that it serves, the book seeks to progress step by step towards the most important question of all—namely, how armed conflict ought to be conducted."[7] Clausewitz might have simplified the text, had he lived to complete it; or perhaps "clarify" is a better word, because it is doubtful he would have succumbed to the temptation to simplify his theory, so set was he against the use of over-arching principles.

Clausewitz's influence on the modern military is pervasive, but relatively recent compared to that of his contemporary Antoine Henri Jomini. Unlike Clausewitz, Jomini set down basic principles to guide decisive action: in his *The Art of War*, he wrote, "All my works go to show the eternal influence of principles, and to demonstrate that operations to be successful must be applications of principles."[8] If Clausewitz's theory represents Newtonian physics tinctured by Kantian philosophy,[9] Jomini's approach is almost strictly Newtonian: cause-and-effect in a very straightforward form. Jomini's dictums grew into the familiar and accepted "Principles of War," which we will consider in depth in Chapter 19.[10] Like Clausewitz, Jomini based his theory on Napoleonic warfare; however, unlike Clausewitz he attempted not to *de*scribe war in its totality but rather to *pre*scribe how to win battles.

Outside of codifying martial principles, Jomini's influence perhaps was most keenly felt on the sea, through the naval theories of Alfred Thayer Mahan. Mahan wrote, "there are certain teachings in the school of history which remain constant, and being, therefore, of universal application, can be elevated to the rank of general principles."[11] His quest to establish the United States as a great maritime power led him to translate such Jominian principles as concentration, operations on interior lines, and central position into naval usage.[12]

In the realm of airpower, we may cite such theorists as Italian General Giulio Douhet, many of whose Jominian ideas seemed disproved by the experience of World War Two but revived in the wake of Gulf War I (i.e., Operation Desert Shield/Desert Storm); British Air Marshal Sir Hugh Trenchard, whose theories of strategic bombing led to disagreement with the US Army Air Force over how to prosecute the air war over Germany; and US General Billy Mitchell, whose inflammatory rhetoric and resultant court-martial were the media for presenting and defending his airpower theories. Though his contributions are not as well known, we may well include British General and one-time Boer commander Jan Christian Smuts, who in an August 1917 report "identified what became the organizational ambition of military aviators everywhere: an independent air service bureaucratically equal to the army and navy." His second report, which "led directly to the founding of the world's first independent air force, Britain's Royal Air Force," was written in April 1918—three months *before* Billy Mitchell commanded the largest air operation of World War One in the St. Mihiel offensive.[13] And we should also include US Air Force Colonel John Boyd, who articulated the famous "OODA Loop" (sometimes referred to as the Boyd Cycle) of Observing the enemy and the situation; Orienting one's self accordingly; Deciding among various options; and Acting. The OODA Loop works very well to describe one-on-one action, from a fistfight to a dogfight,[14] but it was not Colonel Boyd's only contribution to military theory: he also added three orthogonal levels of war—the physical, mental, and moral levels—to the well-known tactical, operational, and strategic levels.[15] We will consider those levels in Chapter 12.

This all-too-brief parade of theorists—neglecting a host of other luminaries from Julius Caesar to B.H. Liddell-Hart—presents at least a small challenge to our earlier assertion that the military generally eschews theory. If theory is so derided in the military, what are we to make of these thinkers whose ideas are considered so timeless and important?

In truth, *theory* is critical—despite the perception that *theorizing* is neither widely practiced nor actively encouraged. It is true that, in the day-to-day activities of organizing, training, equipping and employing armed forces, theory is often neglected in favor of detailed instructions in the form of Operations Plans (OPLANs); Tactics, Techniques, and Procedures (TTPs); and Technical Orders. But as a framework within which technical and tactical proficiency is applied, theory is still a necessary element to military knowledge and understanding war.

10

CATEGORIZING WAR

In *The History of Warfare*, John Keegan compiles insights about how wars have been fought through the ages. One of the most fascinating subjects he examines is how warfare in primitive societies was stylistic and often nonlethal, more like modern boxing or wrestling than modern warfare.[1] This still fits the premise that war must involve violence, since even stylistic war involved violent—though controlled—encounters.

War as a whole, then, can be subdivided into different types with different attributes. We may consider several possible ways of characterizing them, and still leave plenty of room for discussion and debate. Just as the types of war examined by Moseley in *A Philosophy of War* (primitive warfare, political war, modern, nuclear, and post-modern warfare) differ from those Vasquez examined in *The War Puzzle* (wars of rivalry and wars of inequality), the types of war examined here reflect a different approach and purpose.[2]

Primitive war was often but not exclusively representational, in which a few select champions fought each other with the outcome being accepted or at least acquiesced to by the rest of their tribes—think of the Biblical example of the young man David facing Goliath of the Philistines.[3] As the means of killing and the organizational structures of industrial society developed, war became a matter of massed troops arrayed and advancing in choreographed formations. Technological advancements of our day have led to warfare in which a very few troops may wield more killing power than entire ancient armies. At the same time, those technological advances have split warfare into a number of different possible classes: large maneuver-based battles are still possible, but so are small-scale attacks and "cold" standoffs in which shocking amounts of military power are used to threaten but not destroy. Table One compares some of these different types of war, though we recognize that other possibilities may not yet have occurred to us.

Table 1. Some Types of War. Not intended to be all-inclusive.

TYPE	Primitive	Structured	Technological
CHARACTERISTIC	Representational	Massed Troops	Specialized; weapons-dominated
NUMBER OF COMBATANTS	Very Few	Many	Few
LOGISTICS SUPPORT	Very Low	High	High
INTERCONNECTEDNESS	Low	Moderate	High
CENTERS OF GRAVITY	Enemy Soldier	Troops	Nodes of Enemy System
SPEED	Slow	Moderate	Fast
LETHALITY	Low	High	High to Moderate
HISTORY	Staged tribal conflicts, prevalent from antiquity to present-day primitive societies	Rise of armies (Greek, Roman, Chinese) through Napoleonic Era and into 20th Century	Possible to trace to Agincourt; predominantly late 20th Century to current day

TYPE	Cold	Amoebic? Nebulous?	Complex?
CHARACTERISTIC	Weapons-dominated standoff; proxy campaigns	Asymmetrical; hidden enemies	Unpredictable
NUMBER OF COMBATANTS	Few	Few	Few
LOGISTICS SUPPORT	High	Low to Moderate	Moderate? High?
INTERCONNECTEDNESS	Much	Moderate to High	Extensive
CENTERS OF GRAVITY	Weapons; political will	Ideologies; political will	Multiple
SPEED	Slow	Moderate	Fast
LETHALITY	Low	Moderate	Moderate? High?
HISTORY	Nuclear Age	Revolutionaries Crusaders Bellicosi*	Next step after Technological?

*See text for an explanation of this term.

Each of the types of war charted in Table One, and any other type of war the reader might characterize, involves the constituents of war discussed earlier: a military force on one side, an objective, and a military force on the other side. Some types involve higher or lower levels of technical and operational complexity, more or fewer instruments of power, etc., but in each case the military must either threaten violence to (or visit violence upon) an opposing force, military or otherwise, to obtain the objective—or to keep the OPFOR from obtaining its objective.

Using these criteria, the Kosovo campaign of the late 1990s was certainly a war: NATO fought—i.e., applied collective violence, albeit with a good deal of restraint and selectivity—against the indigenous military and paramilitary forces to achieve an objective. The nature of that objective, i.e., unseating a legitimate though reprehensible national leader—Milosevic had been elected President of Serbia (and therefore one of the eight co-leaders of Yugoslavia) in 1989, and assumed the presidency of the Yugoslav Federation in 1997—leaves the action open to criticism as in the sidebar "NATO's War in Kosovo" in Chapter 3, but does not make the action something other than a war.[4] We may apply similar criteria to call the later NATO action in Libya a war, and indeed to call all similar actions "wars."

Since 11 September 2001 we have found ourselves engaged in a Terror War, which under the definition posed above also has the characteristics of war: for us, an aim of continued liberty and self-determination without threat or coercion by outside bodies; opposing forces, albeit nebulous ones; and the use of force. Even though terrorism does not constitute warfare in the classic sense, our response to the attacks across national borders by sub-national groups certainly fits the definition. Thus, in discussing the terminology of terrorism in Chapter 2, and in Table One above, we introduced the term "bellicosi" to represent those international groups that cannot be considered insurgents or revolutionaries, and whose actions have been characterized as "terrorism" even though their motivations do not match those of traditional terrorists.

We may also turn the analysis around and examine whether the characteristics of war apply to the Terror War for the Islamic bellicosi as well. From their perspective it must certainly fulfill all three criteria: the use of force, expressed in bombings and other attacks; an aim of spreading their particularly fascist brand of Islam throughout the world;[5] and opposition, which they locate in the Western world as typified by the US. We should not, however, mistake their momentary focus on the West for the salient characteristic of their war: as much as their rhetoric impugns the US, the rest of Christendom, and especially Israel and the Jewish people, there is no place under any Al Qaeda-taught or Taliban-sponsored sharia society for Hindu, Shinto, Taoist, or any other belief systems. They aim at the West for now, because our wealth and culture have attracted their attention, envy, and ire, but there is no reason to believe that they, like Alexander, would stop before entering the Indian subcontinent or fail eventually to turn their attention farther eastward to China.

One potential brake on the juggernaut of Islamic fundamentalism could be if Muslim nations became competitive in other arenas than the export of hatred and bellicosity, e.g., by production of valuable goods and services. As Francis Fukuyama noted in 1993,

> The current revival of Islamic fundamentalism, touching virtually every country in the world with a substantial Muslim population, can be seen as a response to the failure of Muslim societies generally to maintain their dignity vis-à-vis the non-Muslim West. Under the pressure of competition from a militarily dominant Europe, a number of Islamic countries in the nineteenth and early twentieth centuries undertook crash modernization efforts. . . .[6]

These efforts were not uniformly successful, and many subsequent fundamentalist "reforms" summarily rejected them. But if we substitute "economically" for "militarily" dominant, and place the argument in the current century, what are Islamic countries doing to improve their situations?

Much of the economic power wielded by Islamic countries—especially Arab Islamic countries—is based on natural resources, yet they do not seem to have much incentive to develop their own means of extracting and benefiting from those resources or building robust economies on other bases. Many sheiks seem content to let foreigners do the bulk of the work, perhaps because foreigners first brought the desire and the means to extract the oil from the desert. They accumulate and spend petrodollars when they could be seeking alternate sources of wealth for the coming day when the oil left in the fields is too costly to extract. Perhaps because they come from outside the culture of capitalism, they cannot recognize that they are equivalent to rich young heirs whose only prestige was inherited, and therefore they will never be accorded the same respect under the (capitalist? Protestant? liberal democratic?) work ethic as the poor upstarts who work tirelessly to make their fortunes.

Muslim Spain was a scientific, economic, and cultural wonder; Muslim Indonesia and a few other countries seem to have similar potential, but it is difficult to foresee if their economic successes will stem the tide of the Islamic bellicosi. Unless and until it does, we should not underestimate the Islamofascists as enemies in what certainly seems to qualify as a "war" for them; if nothing else, we should not underestimate their fanaticism.

Moving on, we may be comfortable with categorizing war as a whole or as separate types, and deciding what conflicts fit within what categories. For instance, we may be comfortable with the proposition that the Kosovo campaign be considered a war. We may be less comfortable with largely academic propositions about how wars have, or should be, fought.

For instance, in Chapter 1 we asked the question, "*How* do we know that airpower was or was not decisive in Kosovo in the late 1990s . . . ?" and subsequently we mentioned John Keegan's assertion that airpower *was* decisive. We will not consider in depth whether that conflict represents the first time airpower alone won a war, but let us note a few facts about the larger campaign that bear on that question. It is certain that airpower was used to good effect: the after-action report on the campaign noted that

> NATO generated 78 days of continuous, around-the-clock operations, flew 38,000 sorties with only 2 aircraft failing to return to base, and suffered no combat fatalities. Strike operations achieved effective targeting and weapon effects with relatively low collateral damage.[7]

It is true that no NATO ground force engagements were involved in the actual campaign, beyond an unfortunate incident on the border between Yugoslavia and Macedonia,[8] but equally true that nearly 20,000 KLA troops were involved in fighting throughout the conflict and that 50,000 troops under the aegis of NATO were stationed on the Yugoslav border and moved into the area once the Serbian forces capitulated and left Kosovo.[9] We may never resolve whether that collection of facts proves or disproves Keegan's assertion, however, because the assertion is the result of interpreting facts rather than the facts themselves.

And how we interpret such facts—e.g., how our concepts (or theories) about airpower and conflict influence whether we conclude that airpower was important, or decisive, or sufficient, in a particular engagement—is directly related to the Theory of Knowledge we examined in Chapters 4 and 5. Even when we do not recognize them as such, *all* attempts to categorize warfare constitute at least rudimentary theories. They are abstractions from reality by which we seek to organize the chaos of war and the practice of warfighting. With that as a basis, then, let us examine some specific theories of war.

11

THE LURE OF SIMPLICITY

Our knowledge of any subject grows as we devise new theories and new conceptual formulations and test them against experience. To improve our knowledge of warfare, then, we may combine existing theories or formulate new ones, though we run the risk that our new constructs will be more intricate, rather than more basic, models of war and warfare. We may hope, then, to develop a stronger *coup d'oeil* (as discussed in Chapter 6), because in the midst of such complexity the "stroke of the eye" will become particularly helpful, and perhaps even more critical, to military decision makers.

It may help to consider briefly the difference between analysis and synthesis. Analysis, by breaking large items into their component parts, presents a close-up view; synthesis, on the other hand, unites sometimes disparate items into a larger view. Analysis shows us individual threads, while synthesis shows us the entire tapestry; analysis shows us the individual pixels, while synthesis shows us the entire television or computer image. In developing a cohesive *coup d'oeil* from many different inputs, synthesis may be a greater asset than analysis. (We will return to this idea in Chapter 23.)

Is it any wonder, though, that we prefer the simplest solutions to the most complex problems? Every potential theory that we will examine as we go forward—the levels of war, the Five Rings, the "Principles of War," etc.—simplifies a larger whole. Each may be considered a model that describes part of the whole of warfare, or a lens through which warfare is viewed from a particular vantage point. As simplifications, they subject the complex substance of war to human interpretation, and can be powerful if limited tools.

Unfortunately, in our search for simple solutions, sometimes we may over-simplify. If, for example, an output depends on dozens of inputs, lumping some of the inputs together or weighting them according to influence may make the analysis easier. If we can reduce the number of factors down to a few with the most influence, then the simplification may aid our understanding. Too often, however, we take things too far and reduce the number down to *one* factor we believe is key, leaving out other potential influencers. This is the temptation that caught Jomini and his supporters in their emphasis on "concentration" above any other principle.

At least part of the allure of Jomini's principle of concentration is its inherent simplicity: if I apply the bulk of my forces against a fraction of yours, I am sure to win. Never mind the play of chance, the possibility that I may apply my forces haphazardly or without considering such factors as terrain and weather and morale—the numbers predict success, and success is therefore assured. In Chapter 19, we will examine the principle of "mass" in greater depth and throw some harsh light on the historical utility of Jomini's principle, even though Jomini often qualified it himself by specifying that mass (or concentration) must be applied at the decisive time and place. He might explain away failures of the theory by claiming the commanders had deviated from his true principle; e.g., that they did not concentrate sufficiently, or did so at the wrong time or in the wrong place.

As tempting as such simplifications are, they may cause other unanticipated and unwanted problems, especially if we are tempted to pretend that complexities do not exist. For example, we may assume relationships to be linear when they are not. Or we may decide that interactions between factors need not be considered—yet the way things interact with each other can produce quite powerful results (as we will see in Chapter 22).

As we attempt new formulations, then, we must accept the possibility that we may fail miserably. We may also succeed brilliantly, but most likely we will succeed only in part. As we admitted when discussing the possibility of a "unified theory" of war, the overall effort may ultimately be futile. Attempts to combine vastly dissimilar theories would probably meet with the same dismal results as attempts to combine the philosophies of Aristotle and Plato.[1] So perhaps it is best if we begin with something a little more basic.

12

THE LEVELS OF WAR, AS THEORY

The levels of war—strategic, operational, etc.—are themselves theoretical abstractions; the old Army *Field Manual* 3-0 called them "doctrinal perspectives."[1] The levels are convenient ways to organize our thoughts and, like most theories we encounter, have been refined and expanded over the years.

The word "strategy" itself derives from ancient Greek, but only came into use in the late eighteenth century.[2] Writing approximately a half century later, Clausewitz laid out basic definitions of tactics and strategy —"tactics teaches *the use of armed forces in the engagement*; strategy, *the use of engagements for the object of war*"—which have grown into several modern interpretations.[3] As war grew more complex, the basic levels of tactics and strategy grew to include the operational level and the level of grand, or national, strategy—although it may be said that all the levels were in existence but unnamed and unrecognized. Early on, Jomini wrote of Grand Tactics as opposed to Minor Tactics, and his definition of "strategy" included our current strategic and operational levels of war.[4] Grand tactics at least had the advantage of being understandable from observation and movement, as it dealt with maneuver beyond the range of the enemy's ability to interfere, a concept that became somewhat muddled when grand tactics was subsumed into operations.[5]

It is hard to develop common understanding when we sometimes use the same or similar words to mean divergent things. "Operations" of one type or another can be so separate and distinct that it seems a shame we have not developed particular terminology for them. We can be said to be "operating" if we are participating in a fairly wide variety of activities that may or may not have anything to do with applying military force. The "operational level of war" is something else entirely, however, and might make more sense if it were called "grand tactics." But we will deal with the operational level in depth below.

In discussing the concept of different levels of war, Liddell Hart wrote, "Clear-cut nomenclature is essential to clear thought, and the term 'strategy' is best confined to its literal meaning of 'generalship'—the actual direction of military force. . . ."[6] Eccles provided still another expansion of the definition of strategy beyond its classical interpretation, and one that approached grand strategy: "the comprehensive direction of power to control situations or areas in order to attain broad aims or objectives."[7] And in his 1999 book *Modern Strategy*, historian Colin Gray wrote, "Strategy is the bridge that relates military power to political purpose; it is neither military power *per se* nor political purpose [but] *the use that is made of force and the threat of force for the ends of policy*."[8]

The official US military definitions are refreshingly straightforward, but also problematic in certain ways:

> *strategic level of war*—The level of warfare at which a nation, often as a member of a group of nations, determines national or multinational (alliance or coalition) strategic security objectives and guidance, then develops and uses national resources to achieve those objectives
> *operational level of war*—The level of warfare at which campaigns and major operations are planned, conducted, and sustained to achieve strategic objectives within theaters or other operational areas
> *tactical level of war*—The level of warfare at which battles and engagements are planned and executed to achieve military objectives assigned to tactical units or task forces.[9]

Starting at the top, the definition of the strategic level is closer to what often is referred to as "grand strategy," especially in its inclusion of coalitions and alliances. Moving downward, the operational level seems directly referenced to the strategic in the mention of strategic objectives, and the tactical level is defined almost as a tautology: the tactical level is the level at which tactical units act.

Even naming the various levels has been contentious at times. General Sir Rupert Smith called what we know as the operational level the "theatre" level, "the level that links the tactical to the strategic,"[10] but that seems to make the operational level more of a convenience than a useful concept, at least from a definitional point of view. At one time US doctrine referred to another blended concept, the "theater strategic environment," but that term was dropped from later guidance documents.[11]

Consider, for a moment, the names of the US Air Force's former Strategic and Tactical Air Commands, and the current US Strategic Command. These names were based on the definition—if you will, the theory—that equated levels of war with specific targets and weapons, and in particular the strategic level with nuclear weapons. As Eccles stated in *Military Concepts and Philosophy*,

> One school of thought seems to consider that strategy and destruction are synonymous. This view implies that the greater the destructive power of a weapon, the greater its strategic value. This has the apparent corollary that the availability of a weapon should determine the national strategy.[12]

From the advent of "daylight strategic bombing" in World War Two to the days of our nuclear standoff with the Soviet Union, interpreting "strategic" in that particular way made some amount of sense. In the current international environment, though, we face an array of threats both conventional and nonconventional, and even though large-scale nuclear capabilities still exist, strategy should not be limited to the choice of a single weapon system. As the famed samurai Miyamoto Musashi warned in 1645, "You should not have a favorite weapon."[13]

When SAC and TAC were merged into Air Combat Command (with the SAC refueling arm merged into Air Mobility Command), the US seemed to be shifting "strategic" away from strictly nuclear warfighting, a change that played out a dramatic conclusion on 1 October 2002, when USSTRATCOM assumed responsibilities previously vested in US Space Command. That change, however, appears to have been little more than a consequence of standing up the new US Northern Command to coordinate military participation in homeland defense in the post-9/11 environment. Since the Air Force stood up Air Force Global Strike Command in 2009 as a direct successor to SAC, and then in 2016 resurrected TAC as part of ACC, what seemed to be a shift away from a weapons-oriented approach to the strategic level was perhaps only a pendulum swing.

The strategic level of war has also been identified with an enemy's centers of gravity: those points which, if attacked successfully, may be expected to severely impact the enemy's ability to wage war.[14] At least one writer's analysis, however, appeared to be internally inconsistent. US Air Force Colonel Phillip Meilinger maintained the importance of strategic attacks by stating, "It has long been the goal of military commanders to paralyze an enemy rather than fight him . . ." —which hearkens back to Sun Tzu's observation that defeating one's enemy without fighting is the pinnacle of success[15]—while also having criticized Sun Tzu's admonition to attack the enemy's strategy.[16] Whether this was the result of a misreading of Sun Tzu—assuming his guidance to be reactive rather than proactive—or the legacy of an inferior translation, is unclear.

These shifting interpretations and definitions verify that change is certain, and support our central theme that the levels of war are categories that represent the basic essence of theory: little more than a way of thinking about war and warfare. They do not, however, qualify completely as a theory because they are neither predictive—i.e., they do not provide a sound basis for predicting the outcome of an action—nor are they entirely explanatory.

But if the levels of war, while helpful in organizing our thinking about war, do not rise to the level of theory, then the question remains as to whether they are as helpful as they seem.

For example, consider briefly whether the airpower tenet of "centralized command and decentralized execution" applies evenly across all levels of war.[17] According to this tenet, command of air operations should be centralized under a commander well versed in the use and limitations of airpower, while execution of the various parts of the overall mission is parceled out to the units performing those discrete tasks; they are "the fundamental organizing principles for airpower."[18] However, in practical application "centralized command" seems to apply most readily at the strategic and operational levels: in the Air and Space Operations Center, where the Air Tasking Order is written.

"Decentralized execution," in contrast, is by its very nature a tactical level issue.

As for the operational level itself, it appears to have come to the fore after we realized that concepts suited to fighting nuclear war (if "nuclear" is indeed synonymous with "strategic") and concepts suited to fighting small "tactical" engagements had not been sufficient to conducting war on the scale of Vietnam. The operational level, shoehorned between the strategic and tactical, would provide a conceptual touchstone for explaining, understanding, and perhaps even planning and winning theater-sized campaigns. According to *Joint Publication* 3-0,

> The focus at this level is on the planning and execution of operations using operational art: the cognitive approach by commanders and staffs—supported by their skill, knowledge, and experience—to plan and execute (when required) strategies, campaigns, and operations to organize and employ military capabilities by integrating ends, ways, and available means.[19]

That convoluted description incorporates a definition of "operational art," which is particularly interesting in that it apparently involves planning strategies, when we might have thought that the hierarchy would consist of strategic art planning operations. Nonetheless, a "campaign" is described as "a series of related military operations aimed at achieving strategic and operational objectives within a given time and space,"[20] which seems again to tie together tactics ("military operations") with operations and strategy, though it appears the DoD no longer considers the campaign "the central organizing instrument for joint warfare."[21]

Earlier we called the operational level a convenience, and so it was and is. Was it always extant in the conduct of war, or is the operational level really a new phenomenon, a condition of modern warfare that did not exist for Clausewitz and the classicists? If we take the position that the level always existed but was undiscovered—like a new element on the periodic chart, or a new island in the middle of the ocean—then we may rightly either question its significance or question the wisdom of those who overlooked it for so long. If we accept it as a condition of modern war—that is, if we accept that modern war entails combinations of forces and complex conditions that necessitated a new level concept —then we must admit that supra-modern war (i.e., war of the future) may require developing even further refinements to the concept of levels.

We may accept that as war changes our theories and concepts of war must change to keep pace. Therefore, the operational level makes sense as a concept suited to complex combined arms, perhaps useful in this way: tactics are *how* we perform operations; operations are *what* obtain strategy; and strategy is *why* we pursue tactics in the first place. Even so, this concept of levels of war could stand to be refined to produce a more complete understanding.

And, in fact, the levels have been refined in a certain sense. USAF Colonel John Boyd, creator of the OODA Loop, also proposed three additional levels—though perhaps "components" might be a better word —of war: the physical, mental, and moral levels of war. *MCDP 1, Warfighting,* treats these as "forces" of war in its first chapter—"The Nature of War"—and notes that

> War is characterized by the interaction of physical, moral, and mental forces. . . . Although material factors are more easily quantified, the moral and mental forces exert a greater influence on the nature and outcome of war. . . . Because it is difficult to come to grips with moral and mental forces, it is tempting to exclude them from our study of war. However, any doctrine or theory of war that neglects these factors ignores the greater part of the nature of war.[22]

Why are the mental and moral forces so important? Because they feed directly into the willingness to begin and carry out a war. We can point to various historical cases in which the will was great but the physical capability waned, but we can also find opposite cases in which powerful forces were sapped of their will by protracted, fruitless efforts.

William S. Lind suggested that Boyd's levels may be considered adjacent to the "traditional" levels, such that together they form a grid with the tactical, operational, and strategic levels along one axis and the physical, mental, and moral levels along the orthogonal axis. In 2015, he even included such a nine-box grid in his slender volume, *4th Generation Warfare Handbook*.[23] He had noted previously that "Our armed forces focus on the single box defined by tactical and physical, where we are vastly superior. But non-state forces focus on the strategic and the moral, where they are often stronger, in part because they represent David confronting Goliath."[24] The physical, mental, and moral levels of war—or forces of war, or components of war—are a prime example that it is possible to think of ancient topics like war in new ways.

Do these levels, whether the traditional strategic-operational-tactical triad or the newer moral-mental-physical construct, adequately explain the past and predict the outcome of future decisions, as a good theory should? Not completely. Can any "level of war," past, present, or future, serve such a role? Doubtful. We may interpret an encounter in terms of tactics, operations, or strategies alone; or we may interpret the same encounter in terms only of tactics and strategies, if the operational level was unnecessary; or we may include additional levels and interpret the event accordingly. In each case we may glean some different understanding. But with respect to predictions, these levels may help us categorize a situation such that we may make better predictions, but they do not themselves confer any predictive power.

13

THE MEDIA OF WAR, AS THEORY

In contrast to the levels of war, the media of war are, for the most part, more concrete—after all, they form the physical environment within which war is waged. We stand upon the ground, sail upon and dive under the sea, fly through the air: they exist, they are not abstract notions. We must deal with their reality as we find it, and we cannot wish for it to suddenly be or behave in some fashion that fits our preferences. But does the way we categorize the physical environment of war constitute some sort of theory?

If we expect theories to explain the past, in whole or even in part, then the media of war do not qualify as complete theories in themselves even though they will be important (and may even be dominant) factors in the outcome of engagements. If we expect theories to predict future outcomes with some confidence, then the media of war fail on that account as well. But even if they do not form a useful theory, they should be able to inform the present situations we face.

It may seem self-evident, but amid the continuing emphasis on joint warfare in which media-centric services combine forces, it is important to emphasize that tactical control, i.e., control of a tactical situation, is ultimately based on *presence*. In saying this, we do not mean to detract from joint warfare but to emphasize that we cannot control a medium without a presence in that medium. In practical terms, this means we cannot truly control the air from the ground, the sea from the air, etc.

Recall for a moment the discussion in chapter 10 on whether the 1999 Kosovo campaign was a war and whether airpower alone won it. We noted that NATO's action in Kosovo could be considered a war, i.e., it met the basic criteria; but we presented evidence to challenge whether airpower was sufficient to secure the outcome. It is true that all of the damage inflicted by NATO was done from the air, but it is equally true that a sizeable force waited along the Yugoslav border for the chance to join the battle on the ground. Had the Milosevic government not sought terms, those ground troops would surely have been used. In fact they *were* used immediately after Milosevic sought terms, not as combat troops but as occupying forces.[1] Therefore, despite the opinions of esteemed thinkers on the subject, it seems that the theory that airpower "won" the Kosovo campaign not only remains unproven, but should be viewed with healthy skepticism.

We may even postulate that airpower was not, strictly speaking, *necessary* to the Kosovo campaign: that the NATO forces might have been able to subdue the enemy even without air strikes or air cover. Certainly the probability of success for such a venture was something greater than zero. However, such a hypothesis would be foolish; even if airpower was not *sufficient* to win the war alone, its contribution to the swift conclusion means we can easily consider it necessary in terms of the quick, low-mortality, low-destruction military campaigns we prefer to wage.

It may clarify the point if we stipulate that forces operating primarily from one medium may *affect*, but not *control*, other media. Control implies exploitation, and use for one's own purposes; denying the enemy use of a medium is not automatically the same as using it ourselves. It would be ridiculous to propose that a ground forces commander can, by virtue of having a few antiaircraft artillery batteries or a Patriot missile system, exert control of the air; yet for some reason airmen have at times shown no compunction about claiming that air forces can control the ground. The claim presents itself most fully in the idea of an "air occupation," as typified by the no-fly zones in Iraq under Operations Northern and Southern Watch, although "air intervention" may be a better term; for example, the *Gulf War Air Power Survey* stated, "Air power can hold territory by denying an enemy the ability to seize it, and by denying an enemy the use of his forces. And it can seize territory by controlling access to that territory and movement across it."[2] Upon reflection, though, it must be clear that those statements are only temporarily true; i.e., they are true only while friendly aircraft are operating in the area. And while the "no-fly" operations certainly imposed our will on the Hussein regime with respect to air operations, they did not confer on us control of the cities, deserts, and other areas of Iraq—nor have subsequent air operations resulted in control of other territories.

Our conclusion from this is simple: rather than control, air power can *influence* the ground; likewise, sea power can influence the land; etc. There may be cases such as Kosovo in which that kind of influence is particularly great, but influence is not control.

Why is this important to our study of theory? We have already conceded that the concept of "media of war" does not constitute a complete, coherent theory. But if the concept is helpful to our understanding of war and warfare then we should be able to draw meaningful conclusions from it that transcend specific tactical examples.

To reiterate: Clearly the media of war do not adequately explain the past, as with the argument over how the Kosovo campaign was won. Questions such as that may never be resolved satisfactorily, if for no other reason than parochialism. Similarly, the media of war offer no real predictive capability.

We can extend the "media of war" analogy, though, and by doing so extend our understanding of war to new environs. For example, can our observations about the traditional media apply to those realms we have not yet mastered, and possibly prepare us to solve the problems those media present? It seems that the idea that presence is required for control must apply across more than the land, sea, and air boundaries. No earth-bound military will ever exert real control over space, for example, if for no other reason than a terrestrial force only sees a segment of the sky at a time. Our worldwide system of networked ground stations achieves "virtual" control of space in nearly every direction, but this has not reached the level of true space "control" yet. For one thing, many orbiting spacecraft at times move in and out of range of ground stations, and must operate autonomously at different points in their orbits; for another, the ability to exert influence into space is limited to those platforms at our command, over links that are as tenuous as any radio signal susceptible to interference. This argues for space-based control, but the problem is further complicated by the vastness of space itself: even a space-based presence will not be able to control much more than its immediate area.

By this logic, some might say that airpower achieved "virtual" (or "de facto") tactical control of the ground situation in Kosovo. The idea of virtual control may apply better to air operations than space operations, since aircraft control their surroundings better because they have the advantage of the flexibility of powered flight. But without continued presence as later provided by the occupying ground forces, airpower's control of the Kosovar ground situation was, if not virtual, at least transitory.

We may be able to do more with the media of war, however, than simply acknowledging them as the environs of war. Before we work on that possibility, let us consider new media in more depth.

14

MORE MEDIA OF WAR: THE POSSIBILITY OF SPACE

Even though no battles have been fought there, we may consider space to be a fourth medium of operations based on the presence of military hardware in it. Despite the short-lived implementation of many of the turn-of-the-21st-century "Space Commission" recommendations, however, we have not yet established a cogent working theory with respect to space.[1] Partly, this is due to our lack of experience there. At this writing it is just under sixty-five years since the US Air Force established its Western Development Division to design and build ballistic missiles and then satellites, and likewise just over sixty years since Sputnik was launched; but, as USAF Major Shawn Rife pointed out in a 1999 article, "we still cannot operate in space nearly as easily or routinely as air forces could operate within a decade of the Wright brothers' first flight."[2]

The historical development of sea power is an apt analogy to our current space capabilities. With our focus on Earth-orbiting satellites and weapons that only transit space, we seem to be at the same point as our ancestors were when they plied coastal waters but did not range across the oceans.[3] If that is the correct historical analog, it is no surprise that we have no strong space power theory yet: coastal sailors likely did not spend much time theorizing about deep ocean warfare.

Another fact may help illustrate the current difficulties of space operations: namely, that spacecraft follow fixed orbital paths, and it is costly and time-consuming to deviate from them. This restriction is similar to the dependence on railways during and after the US Civil War when armies "became tied to the railhead."[4] Even if this is closer to reality than the coastal waters analogy, cogent space power theory may still be a long way off, and space operations likely will not come into their own until technology overcomes the orbital limitations. Consider the difference that Liddell Hart described between railroads (and motorways) and tanks—

> The greatest development which the tank had produced in warfare—more important even than its revival of the use of armour—was its ability to move off the road, to be free from dependence on the smooth and hard surface of a prepared track. Whereas the wheeled motor-vehicle merely accelerated the pace of marching, and reproduced the effect of the railway in a rather more flexible form, the tank had revolutionised mobility. By laying its own track as it moved it dispensed with the need to follow a fixed route along a prefabricated track, and thus superseded one-dimensional by two-dimensional movement.[5]

—and imagine the difference in theory and application between fixed orbits and powered exoatmospheric flight.

How should we evaluate a space power theory when it emerges? A good theory of space warfare should explain the success or failure of past encounters and predict the outcome of future encounters. The task of would-be theorists at this time should be to formulate possible theories that will be tested in the conflicts that are certain to come. That is somewhat difficult at present because, even though "space" has become more widely accepted as a fourth medium of war, that acceptance is not yet universal (no pun intended).

The 1999 Department of Defense Space Policy stated that, "Space is a medium like the land, sea, and air within which military activities will be conducted to achieve U.S. national security interests," but the latest version of that policy no longer includes that statement.[6] Some commentators refute the possibility (e.g., ". . . the chances that [space] can be used as a war-fighting arena, in and of itself, are slight . . ."[7]), sounding for all the world like the pundits who predicted that powered atmospheric flight was impossible, let alone useful for war. Others state, "In order for space power to reach its full potential, . . . space must be recognized as a geographical environment for conflict that is, in a strategic sense, no different from the land, sea, air, and the electromagnetic spectrum."[8] In the US Air Force, service doctrine went from considering space as a portion of the "aerospace" domain to giving it full status as the service claimed to be an "air and space" force—then the pendulum swung back toward "aerospace," a single continuum—before swinging back to the idea of two separate (but equal?) and connected media.[9]

The on-again, off-again treatment of air and space as one continuous "aerospace" medium would have us believe that altitude is the prime factor in defining an environment, as in the 2015 Air Force *Basic Doctrine*: "Airmen exploit the third dimension, which consists of the entire expanse above the earth's surface. Its lower limit is the earth's surface (land or water), and the upper limit reaches toward infinity. This third dimension consists of the air and space domains."[10] But using the vertical dimension as a determining factor in this way is fundamentally weak, as it merely restates the "aerospace" idea without using the word. It treats air and space as physically separate while considering them doctrinally indivisible: a fuzzy concept, and one that does not bear up under much scrutiny.

Proponents of the "aerospace" idea did not, for instance, equate "aero" with "air"—following, for example, the British "aeroplane" which is synonymous with "airplane"—in which case "aerospace" would mean the same as "airspace." Nor did they acknowledge that *all* military forces operate in three dimensions; even Sun Tzu expressed the ages-old preference for operating from the high ground.[11] They also did not advocate blending other media—e.g., land and sea. Neglecting the possibility of true underground warfare (which we may consider the "low ground" and has some limited effectiveness, as in the Viet Cong's warrens and Al Qaeda's caves and tunnel complexes), the "aerospace" idea seemed closer to that of surface and undersea warfare, both of which are the purview of the Navy. No one would assign the Navy complete rights to or sovereignty over the "aqueous" environment, if that were defined as every fluid medium containing appreciable amounts of water, because that would include the atmosphere as well—although that claim might be considered to be already well developed in practical terms in the form of aircraft carriers. (See Figure 2.) Yet thinking of the "aqueous" realm is not much wilder than positing the "operational" but not "physical" realm of "aerospace," "verticality," or the "third dimension."

The key is that air and space are not the same—they are two separate physical realms joined together in the abstract, and in fact are fundamentally different—while both surface and undersea operations take place in the same medium. (See Professor Colin Gray's *Modern Strategy* for a detailed discussion of the "grammar of war" as it applies to "terrestrial" operations versus the "vertical dimension."[12])

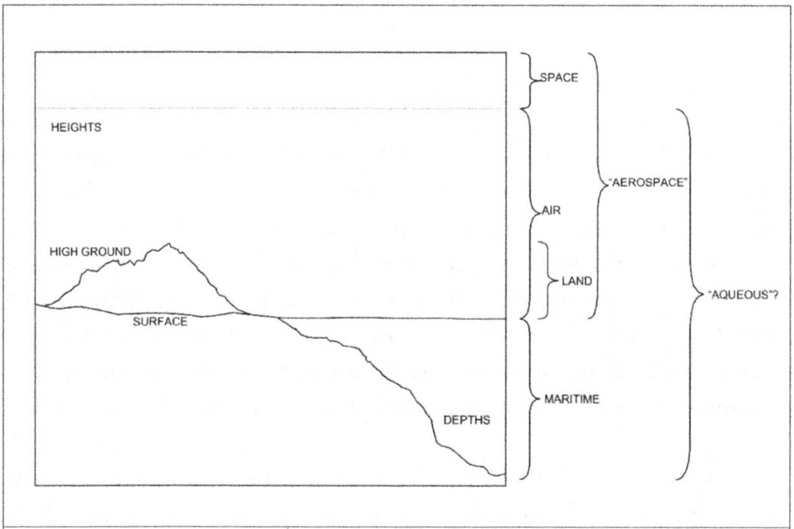

Figure 2. The vertical dimension. All forces operate in three dimensions (or four, if including "time"). Here the dimension is divided into physical, and arbitrary, realms.

Despite its inability to describe operational reality, the "aerospace" idea was hard to shake and still has some support, proving again that "every theory is correct—in its own world."[13] In his book *The Icarus Syndrome*, RAND analyst Carl H. Builder defended the use of "aerospace"—and specifically the idea of an "aerospace continuum"—as a means of giving space equal footing with air. He wrote, "Space has been a stepchild too long."[14] The kind of inequities he noted were also pointed out by the Space Commission, which emphasized the separate nature of air and space rather than the notional combined medium.

All the wrangling over terminology and the switches to and from "aerospace" as doctrine have, from a distance, seemed primarily political (as did the recent and surprising Presidential call to work toward establishing a separate "Space Force"). Unfortunately, they have also at times resulted in theories that contradicted physical reality, since no aircraft in the inventory can operate in space; no spacecraft can operate in the atmosphere; and the physics of powered flight versus orbital mechanics are vastly different.[15] The fact remains that the presence of atomic oxygen, charged particles, and even particulate matter in the practical vacuum of near-earth space does not mean it can be treated or used as a physical atmosphere, any more than the fact that seemingly solid objects are, at the atomic level, actually mostly empty space means they can be treated as vacuum.

A pilot dropping a bomb must take into account an astounding array of factors before release—altitude, airspeed, weather, wind, etc.—a process made somewhat simpler by electronic aids and weapons, but still impressive. A space warrior (eventually) dropping a weapon from orbit must contend with similar, but not exactly the same, factors—orbital height (which determines orbital velocity), inclination, eccentricity, the curved orbital transfer required to place the weapon in the thickening atmosphere, drag, friction heating, weather, wind, and so forth. Until we create machines capable of transiting to and from space with ease, only the rudiments of airpower theory and practice can apply to the possibility of war from space—and the possibility of war *in* space, however remote, requires a theory all its own.

Treating space as a medium of war, or choosing to treat it as an adjunct to atmospheric operations, are both conceptualizations meant to categorize and aid understanding. Between the two, it seems keeping air and space separate and distinct is more likely to explain past events and predict successfully future outcomes; therefore, space as a medium unto itself appears to be a more powerful (though still not complete) theory than the "aerospace" construct.

We may ask whether electronic signals (e.g., communications signals, information from missile warning satellites, etc.) may work as well under the aerospace idea as the air-and-space notion, since they are less affected by gravity and weather than items of physical bulk.[16] Certainly a case may be made, especially when those signals originate on geosynchronous platforms or GPS-style constellations with near-constant coverage; however, use of signals from other orbiting assets (e.g., weather satellites) requires knowing those satellites' orbital characteristics—which are best explained by space-only terms. Thus 2003 Air Force doctrine stated,

> Air and space are separate domains requiring exploitation of different sets of physical laws to operate in, but are linked by the effects they can produce together. By using the phrase "air and space" instead of "aerospace" we acknowledge the inherent differences in the two media and the associated technical and policy related realities without deviating from our vision.[17]

Later Air Force doctrine made the same claim about "air, space, and cyberspace," thus wrapping electromagnetic media and "virtual" spaces together with the other two domains.

More interestingly, the 2011 version of the Air Force's capstone doctrine airpower "as a unitary construct"—a definitive "change from previous basic doctrine which talked to 'air and space power'" that was clearly intended to "use concepts and language that bind Airmen together instead of presenting the Air Force as a collection of tribes broken out in technological stovepipes according to the domains of air, space, and cyberspace."[18] (The same "unitary construct" explanation, minus the bit that differentiated it from previous doctrine, was in the 2015 *Basic Doctrine.*) Even so, the *Basic Doctrine* noted the separation of the physical domains:

> From an operational perspective, the air domain can be described as that region above the earth's surface in which aerodynamics generally govern the planning and conduct of military operations, while the space domain can be described as that region above the earth's surface in which astrodynamics generally govern the planning and conduct of military operations.[19]

Before we consider the possibility of cyberspace in more depth in the next chapter, let us briefly consider the "near-space" region that gained interest in military circles around 2005. This region of the atmosphere between 65,000 and around 300,000 feet (about 100 km) in altitude could support long-duration airships—either lighter-than-air craft or "high altitude long endurance" uninhabited air vehicles—that would perform surveillance and communications relay functions.[20] By the time the military began seriously considering near-space, experiments with balloon-borne radio relays had already been accomplished and commercial firms had been using them to monitor widely spaced ground installations for several years.[21] How long it might take to prove the military utility of this realm is unclear; however, at this stage it seems safe to observe that near-space is not space because aerodynamics is still the governing factor, as opposed to orbital mechanics that obtain in space. To some that may seem evident, but to others (especially those concerned with who might have operational control of such assets) the idea of "close to space" seems to equate to space itself. The opposing rationale is simple: even though the altitude is too high for sustained lift over a wing, the governing forces—thrust to counteract drag, lift to counteract gravity—are the same as for atmospheric flight and not the same as the governing force of orbital mechanics: i.e., gravity itself.

15

STILL MORE MEDIA OF WAR: THE POSSIBILITY OF CYBERSPACE

Before the bellicosi struck on 11 September 2001, cyberspace seemed to be shaping up—though that cannot be the right phrase—to be the next battleground. Even as the Terror War progressed, and then as leaders pulled back from pursuing it formally, we still looked at network-centric architectures and operations as ways to extend the advantages of our troops and commanders, and to project disruptive power electronically.

"Network-centric" anything requires a network and connections to it, and the proliferation of wireless portable data assistants and other devices makes that possible to a greater degree than ever before. An unintended consequence of this development—i.e., of the ability of General Officers to use smartphones, etc., to participate in the decision-making process wherever they are—may be the reduction of delegation. In 2007, US Joint Doctrine stated that, "These technological advances increase the potential for superiors, once focused solely on the strategic and operational decision making, to assert themselves at the tactical level," as if that was automatically preferable to allowing subordinate commanders to exercise their autonomous authority. The Joint Publication did remind commanders that, "While [asserting themselves at the tactical level] will be their prerogative, decentralized execution remains a basic [command and control] tenet of joint operations,"[1] but it left the door open to a dangerous tendency: not necessarily to micromanage but to perpetumanage, i.e., to manage incessantly.

Thankfully, those two sentences were removed in the latest version of the doctrine. I do not know how they came to be inserted; they should have been abhorrent to anyone interested in developing competent leaders over time. Short-term, that approach might have kept senior leaders available to make decisions when needed. Long-term, however, a generation of up-and-coming leaders may not have gotten the decision-making experience their predecessors had. Watching the decision-making process and even making inputs to it, while they have their advantages, are not the same as participating in it as the decision maker.

Possibly the doctrine writers became concerned when the proliferation of electronic tools led to frequent video-conferencing between the front-line and rear echelon Headquarters. This type of "leading from the rear" can be paralytic when the reachback goes all the way to the seat of power; before the days of videoteleconferences, this weakness was evident in the German Army during World War Two and the US forces in Vietnam. It is no surprise that intrusive leadership from the rear has produced paralysis near the front. Very few commanders in the field have ever been willing—either from lack of faith or lack of will—to follow Sun Tzu in his assertion that "There are occasions when the commands of the sovereign need not be obeyed."[2] Furthermore, while the videoconference may be useful in planning, modern engagements are often short and require very quick decisions; if this temporal compression continues to increase, electronic reachback may not be to our advantage. (Recall the quickly iconic image of President Barack Obama and his top advisors apparently watching the action associated with the execution of Osama bin Laden. It is instructive that the photograph elicited little such concern among commentators; we may wonder how that operation would have fared if the observers halfway around the world had begun issuing orders over the communication links.)

Information is vital to decision-making, irrespective of the means of acquiring and transmitting it. This brings up an interesting question: whether "information" itself may be considered as a fifth medium of war after land, sea, air, and space. Joint doctrine treats "cyberspace" as something like a medium, defining it as "a global domain within the information environment," which it defines as the "aggregate of individuals, organizations, and systems that collect, process, disseminate, or act on information."[3] Referring to information as an "environment" comes rather close to conceiving of it as a medium of war unto itself, which may, for instance, make it easier to consider vulnerabilities, attacks, and defenses. This is good for info-warriors, but what does it mean for others? If the definition of "media" changes because we use it in a new way, does our understanding of some particular media suffer? Or is it improved?

To the immediate point: Is information a thing? Is it a place? Is it both, or neither?

In the sense that it may be produced, stored, accessed, withheld, etc., information has some of the characteristics of things. But what is actually produced, stored, etc., usually is not the information but some collection of symbols in or on a medium. The symbols convey the information but are not themselves the information, for two reasons. First, the same symbols (e.g., the 26 letters of the English alphabet, the ten Arabic numerals, and various punctuations) may be used to convey different information when arranged in specific ways. Second, the same information may be conveyed by different symbols (from pictographs to other languages) although some meaning will be changed or lost in the process. (From a military or intelligence standpoint, the information may also be deliberately hidden in the symbols by way of encryption.)

In the real, everyday world—i.e., neglecting those aspects of theoretical physics and the quantum world that we cannot perceive—land, sea, air, and space cannot fit into that description of information. It may help if we consider the constituent parts of the media of war. Land is composed of, well, dirt; soldiers are not called "ground pounders" for nothing. The sea is composed of water. Air is composed of a mixture of atmospheric gases, of which the largest proportion is comprised of nitrogen and oxygen (with appreciable amounts of water vapor). Space is composed of nothing *per se*—outer space is a region of vacuum and radiation, with greater and lesser concentrations of gasses and energetic particles—but still it can be considered an actual place where clear physical laws apply. And information is composed of . . . what? Data?

No physical analogy holds if we consider information itself as a medium. (Something along the lines of the particle-wave duality of light might apply to information, but let us leave the potential subject of information physics to others.) However, if we consider information as a constituent part of something greater we can at least wrap our minds around the idea of a possible medium: the virtual reality of cyberspace which has been adopted as the realm of US Cyber Command (a component of US Strategic Command). Alternately, and much more concrete, the medium would be not the information or a virtual realm but rather the infrastructure—either physical or algorithmic—that makes up the information network.

It is not clear how to resolve this to produce a useful theoretical base.

Data, when collected and collated and analyzed, may become information. ("May" because they may also be discarded as superfluous or meaningless or outdated.) Information, interpreted by applying theory, may become knowledge; and knowledge, applied over time, may become wisdom. Considered in another way, in light of Theory of Knowledge:

- *data are raw inputs*
- *information is fact-based and relatively static*
- *knowledge is dynamic and useful for prediction*

We might consider what "management" means in relation to each of the above items: data management, information management . . . knowledge management? Wisdom management? Management would generally refer to how each is produced, stored, and distributed, and emphasizes doing so in the right time, amounts, and contexts. But data management, even when terabytes of data are concerned, is far easier than the other two; and from a military viewpoint this again leads us to the electronic storage and transmission systems both as potential vulnerabilities and potential targets.

It is no wonder then that network-dependent militaries wrestle with how to categorize and understand information warfare, and whether it represents a medium unto itself. The prevalence of linked computer networks with research and development information or deployment databases, protected behind firewalls of varying effectiveness, as well as the availability of open-source intelligence information on the Internet, offers a target-rich environment for both friendly and enemy forces. It also offers the opportunity for considering wildly esoteric questions. For example: If "reality" exists in "space," does "virtual reality" exist in "cyberspace"? Would it be true to say that virtual reality is to cyberspace what reality is to three-dimensional space? Granted that cyberspace is an artificiality, as it exists only inside an artifice (i.e., the computer or network of computers), can it be considered a fifth medium of war? And if the space we see around us consists of the three-dimensional world with which we are familiar, what are the "dimensions" of cyberspace?

Furthermore, if we note that physical reality actually has four dimensions—the three spatial dimensions, plus time—does cyberspace extend into the past or the future through mathematical models that computers can run over and over in moments?

We can say little more about information or cyberspace as theoretical constructs. So far as they help us understand and explain the processes of gathering and assessing knowledge about an enemy, they may be useful. Further, as we postulate what may happen and predict outcomes of certain actions, information as a thing and cyberspace as a medium can be useful. As with space, however, the theory by which they are understood is not yet complete.

Thankfully, we need not address all these metaphysical ("metavirtual"?) concerns in order to think seriously about the arena of electronic and information warfare. Today's information warriors fight to protect networks from intrusion, and presumably are also busy developing electronic "smart bombs" to attack specific targets, or growing electronic viruses that, hopefully, will not affect friendly forces the way chemical weapons sometimes do—for example, the "stuxnet" worm that appeared to be targeted against Iranian uranium centrifuges. Many may be concentrating on those tasks without considering the larger questions of how warfare in this arena is alike and different from traditional conflict. Questions like this are important, however, and cannot be ignored for long, if this new breed of techno-warriors is to develop theories and doctrines to guide the application of these new weapons.

16

COMBINING THE LEVELS AND MEDIA OF WAR

As discussed in Chapters 12 and 13, neither the levels of war nor the media of war rise to the expectations of theory: they are informative, in terms of helping us understand the present; and they are enlightening, in terms of helping us place additional factors in context; but neither really is explanatory or predictive. Separating the media of war is particularly useful in placing bounds on the arena, as it were, and dividing the field and flow of battle into different levels is meant to help us understand how battles are arranged and fought.[1] This accords with the purpose of theory, even if the result is useful more for understanding than action.

The demarcations between the traditional levels of war—tactical, operational, strategic—are neither smooth nor continuous, because they are artificial constructs rather than absolutes. As discussed previously, the definitions and descriptions of the levels have changed over time. The levels are most like theories in this regard, evolving to fit greater understanding of the reality they describe. Likewise, the orthogonal physical, mental, and moral levels may also evolve.

The media of war are much more concrete than the levels of war. Land, water, and air all lend to battle their own unique characteristics due to their unique natures, and in Chapter 14, we discussed the new frontier of the fourth medium—the vacuum of space—which poses its own interesting challenges as we come to grips with its promise and its peril.[2] Our strict adherence to doctrines applied to each medium, however, is not absolute. The services have at times worked together well, and in recent conflicts joint action has been quite commonplace, but the Marines are the only service that has succeeded at developing a warfighting approach that specifically addresses each traditional medium:

> The Marine Corps is a suitable example of a *composite* force. . . . Their doctrine writers, their developers of manuals and training, their exercise managers, their budgeteers, their acquisitors, their combat support and combat service support people are all devoted to directing their efforts to the point of conflict. . . . We might argue the wisdom of allowing the Marines to maintain a semiautonomous doctrine, but high marks must be awarded their *compositeness.*[3]

As we consider the broad applicability of theory, and search for theoretical constructs that connect different theories, it seems curious that the levels and media of war have always been treated separately. We have yet to combine them into any coherent framework to understand their interactions. Whether doing so provides any new insight remains to be seen.

At the tactical level, as discussed in Chapter 12, the four media remain separate: land forces primarily contend with enemy land forces, air forces primarily fly and fight against other air forces, etc. (space being the current exception).[4] Joint or cooperative missions, however, that support other services, from air attacks on advancing troops to naval missiles directed at ground targets—from special forces troops taking out airfields to space-based sensors detecting tactical missile launches, and so forth—may be considered to entail the higher levels of operations and strategy (see Figure 3). Combined operations on a theater-wide scale give even bigger opportunities for the media to be considered *in toto*. This is not always easy, however.

It is natural for us to approach problems from the perspective of what we know—i.e., from the perspective of our experience and our theories about the world and how it works—but we might do well to rise above our parochialism if we can. For example, despite significant contrary evidence, "aerospace" power advocates have occasionally claimed that what is predominantly airpower is capable of doing almost all (if not absolutely all) things military. In an article defending the Air Force's "organizational commitment to change the way we think about air and space power," two senior Air Force officers studiously ignored the contributions of other services to an imagined future conflict. They postulated that it might become important to take out an enemy's satellite ground control station, then proceeded to list only "air and space" options for accomplishing the mission.[5] But what about naval or land-based options? To add specifics to their general example, imagine that it became necessary to eliminate the European Space Agency's Maspalomas ground station, part of their ESTRACK network. Located as it is near the shore on Gran Canaria Island, we can postulate several non-Air Force-related methods of destroying it, among which might be offshore shelling, targeting with a Tomahawk Land-Attack Missile, or infiltration by SEALs or Special Forces. Yes, airpower could do the job, but we need not rely on it alone; depending on the circumstances, other methods might be preferable. Focusing too tightly on one realm blinds us to other possibilities.

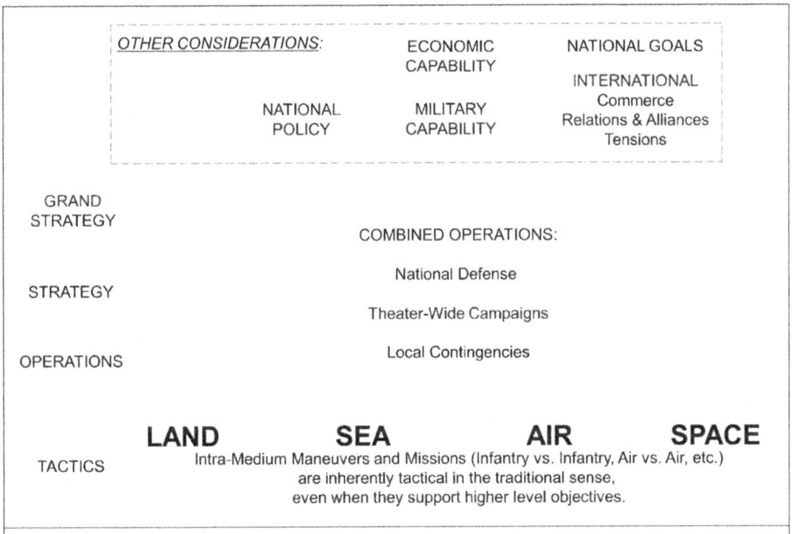

Figure 3. Combined theories? Examining the levels and media of war, looking for relationships.

In contrast, under the construct in Figure 3, media-based forces working together reach higher levels of war than when they remain confined to their individual, natural areas of operations. If we carry this thought further, it leads to a new way of thinking about strategic issues.

Looking at strategy as it relates to the media of war is different from the traditional outlooks discussed in Chapter 12, which ranged from "the use of engagements for the object of war" to operations against the enemy's ability to prosecute the war. Synthesizing the levels and media of war reserves the strategic level for combined operations: i.e., those necessary to prosecute a comprehensive campaign on the order of the Allied campaign against Nazi Germany, or, to a lesser extent, the two Gulf Wars. The plans and programs to support our complete national defense effort would remain at the strategic level, but that level would no longer be the province of particular enemies or weapons or targets (Figure 4).[6]

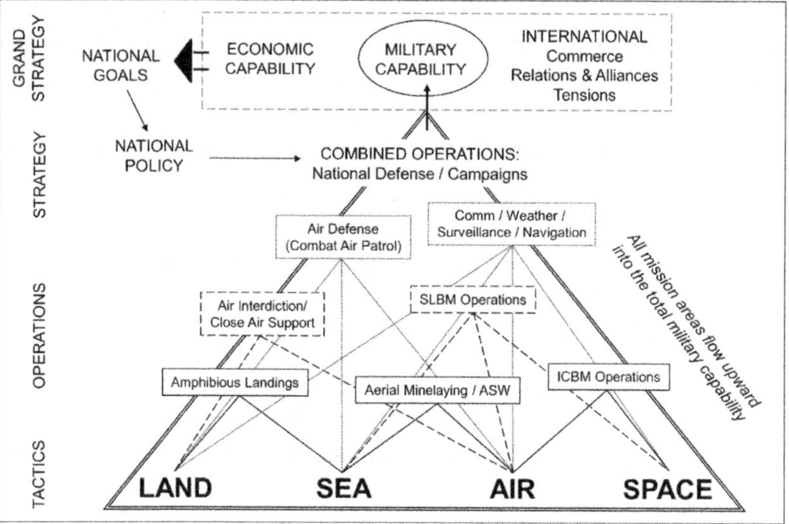

Figure 4. Combined theories? Levels and media of war, showing inter-media missions at the operational level (*not all inclusive*).

This construct, then, reserves the operational level for intra-theater rather than campaign activities.[7] Several new things derive from this combination. For example, if we apply this interpretation strictly, and focus on how the media involved interact, major nuclear weapons such as ballistic missiles fall into the category of operational, rather than strategic, systems.[8] In addition, many systems that play important support roles—e.g., communications and navigation systems—are then seen as strategic assets because they operate across the entire set of media.

Above all of the traditional levels is the grand strategic level, a concept widely recognized abroad but one that has never gotten much attention in the US. In 2004 former Senator Gary Hart wrote, "Lacking a grand strategy, the United States is today seen by too many peoples of the world as a great power without purpose—a giant that pronounces rather than listens—and thus as a great danger."[9] But what is grand strategy? As Liddell Hart stated, "the role of grand strategy—higher strategy—is to co-ordinate and direct all the resources of a nation, or band of nations, towards the attainment of the political object of the war —the goal defined by fundamental policy."[10]

Grand strategy looks beyond the immediate conflict and tries to lay the foundation for the peace that will eventually materialize.[11] Grand strategy combines all of the instruments of power at the government's disposal: diplomatic, informational, military, and economic. Referring to the combined model we have built, the medium of space turns out to be most supportive of grand strategic aims, because space systems support more than the military instrument of power. Military, civil, and commercial satellites provide the space-based infrastructure for all the instruments: diplomatic as well as military communications; timing signals that facilitate economic transactions; and vast amounts of other information, like meteorological and remote sensing data, that inform grand strategic planning.[12]

Does having a grand, or national, strategy—one that focuses all of the military, economic, etc., instruments of power on a set of national goals—mean setting ourselves up in contention with the rest of the world? It need not do so, and some have argued that not having such a strategy has left us in that position anyway.[13] How we articulated any grand strategy would determine whether we were at odds with or in concert with our world neighbors.

Consider that for years Soviet doctrine was explicitly dedicated to the eventual triumph of world Communism, a goal that naturally alienated those countries that preferred not to participate in the Marxist experiment. If we developed a strategy that called for US dominance at the expense of the rest of the world, we could expect to encounter similar resistance, if not outright aggression. And perception is a key factor in this discussion: While we have not explicitly stated a goal of world dominance, our politico-military actions (e.g., using NATO and/or the UN as a front to involve ourselves in Kosovo, and Libya, and Syria, and elsewhere) and even our commercial activities (e.g., exporting US culture through film and other media) may be perceived as dedicated to that aim. Whether these outcomes were intentional or not, when other countries and cultures resist them we encounter aggression we neither fully understand nor appreciate.

That is, we see ourselves as good and our intentions benevolent, but other countries—especially those who may lose out in some way or those who contend with us in some arena—see us quite differently. This is one reason Yugoslavia accused the US of war crimes for acting without United Nations authorization, a claim that had some legal merit but lacked something in moral authority; likewise the sometimes vehement opposition to US policies and actions in other parts of the world. The discrepancy between our actions and their perceived intent or result is often the excuse made for much bellicosi activity—and one reason why the moral force affects the nature of conflicts more than the mental or physical forces[14]—though it seems the leaders of the Islamic bellicosi have an agenda similar to the Soviets: specifically, the eventual triumph of world Islam and the establishment of a global caliphate.

It may be that we are blinded by our own good intentions, and fail to consider that what is good for us may not be good for other countries or other cultures. We must not linger too long in that spotlight, however; some amount of critical introspection may be valuable, but we should not succumb to the fallacy of blaming ourselves for all the world's ills or for other peoples' treacheries. We should recall, especially in dealing with other nations, that the actions of other countries will naturally be more in their interest than in ours, and not automatically succumb to their pressures where the benefit is not clear.

In many cases the course we pursue would benefit not only ourselves but many other countries as well, not (or not entirely) because we are altruistic but because our reliance upon free commerce naturally tends toward mutually beneficial exchanges for the participants—a basic tenet of economics. But the fact remains that we have no overarching grand strategy to guide or inform our national actions.[15]

Regardless of whether we eventually develop a grand strategy at the pinnacle of where the levels and media of war interact, the synthetic view presented in Figure 4 paints a new picture but is unlikely to supplant traditional viewpoints on the levels or the media; perhaps deservedly so. For instance, this model does not directly address irregular or "Fourth Generation" warfare (though Joint Doctrine holds that "Warfare is a unified whole, incorporating all of its aspects together, traditional and irregular"[16]), nor does it strictly incorporate the physical, mental, and moral forces. Nevertheless, it does present a new view of what are often static, if not stale, concepts.

17

QUASI-THEORIES

In the search for simple guidelines to the complexity of war, we often grasp rules of thumb and treat them as if they were well-thought-out theories. For example, the commonly accepted principle of mass championed by Jomini—which we will examine in depth in Chapter 19—is not stated categorically, but appears with caveats: "almost always" a larger force beats the smaller; "usually" a larger force suffers less.[1] Stating it with these qualifications nearly disqualifies it as a principle, although purists will certainly find their way around that conclusion.

What other approaches to explaining how the world works might we consider "almost theories"?

Fog and Friction. Fog is the "play of chance and probability" on the field of combat.[2] As Clausewitz observed, "War is the realm of uncertainty; three quarters of the factors on which action in war is based are wrapped in a fog of greater or lesser uncertainty."[3] Whether three fourths, seven eighths, or some other fraction is immaterial; fog, as a concept, may be considered a veil that clouds operations.[4]

Fog's handmaiden is friction. Friction "refers to uncertainties, errors, accidents, technical difficulties, the unforeseen, and to their effect on decisions, morale, and actions. . . ."[5] This concept recognizes that "countless minor incidents—the kind you can never really foresee—combine to lower the general level of performance, so that one always falls far short of the intended goal"; indeed, Clausewitz asserts that friction "is the only concept that more or less corresponds to the factors that distinguish real war from war on paper."[6]

As seen in the previous discussion regarding the levels of war, the development of ideas and theories to explain reality is often quite dynamic. The concepts of fog and friction have not been redefined in the way the levels of war have been, but they did seem to fall out of favor in the mid-1990s. As Colonel John Warden wrote in 1995, "Napoleon and Clausewitz were right when they talked about friction, fog, and morale. They were right, however, in a time when communications were almost nonexistent. . . ."[7] He continued:

> [Weapons] had little more range or accuracy than those of the Roman legions, most movement was at a walking pace, battles were won or lost depending on the outcome of tens of thousands of almost personal encounters between soldiers who could see each other when they fired, and war was largely confined to the clash of men or ships at a limited point in time and space.

By this Colonel Warden implied that the conditions of war had changed enough that in this modern era fog and friction may be, if not actually discounted, at least downplayed. We cannot deny that warfighting has changed drastically; however, no known changes in weaponry or intelligence-gathering techniques have eliminated or will eliminate the presence of uncertainty (fog) and compounded nuisances (friction). As John Keegan pointed out in *A History of Warfare*, "The most important limitations of warfare . . . have always lain beyond the will or power of man to command."[8]

Colonel Warden seemed to propose that as information systems improve, battle preparedness and battlespace awareness will improve to such a degree that they completely mitigate the fog and friction of war. Other factors of modern warfare work against this proposition. Besides the danger of information overload, we rely so heavily on advanced technology that if the technologies can be negated, fog and friction loom even larger.[9] An obvious example is our reliance on the Global Positioning System for navigation, but the problem goes even deeper. A World War Two infantryman lamented the modern military's dependence on electronics for the simplest of things by saying, "I carried my orderly room in my pocket. I had the duty roster, the sick roster, and on payday I had the pay roster. . . . If you lose power, what are you going to do, plug your laptop into a tree?"[10] Thus Colonel Warden's assertion actually took us backward from modern interpretations of how the world works—including natural variation and chaos theory, both of which recognize that a degree of randomness is present in the outcomes of all events—to a fully deterministic, Newtonian interpretation that cannot hold up to reality.

The concepts of fog and friction predict that plans will sometimes fail for unforeseen reasons, and explain that similar encounters may produce vastly dissimilar results; this lends some credence to the viewpoint that fog and friction represent a theory. However, fog and friction appear powerless to predict what plans will fail or why we achieve different results; therefore, they are more useful to the conduct of war as warnings—i.e., factors which a commander must beware—than as tools the commander can use to increase the probability of success.

Five Rings. The seventeenth century Japanese warrior Miyamoto Musashi wrote a classic text on strategy entitled *Go Rin No Sho* (*A Book of Five Rings*), in which he outlined the tenets of strategy as practiced by samurai. The "five rings" of this section refers not to his text but to the model developed by the aforementioned Colonel Warden to guide the air campaign in the first Gulf War. This modern five-ring theory has not reached the level of eminence of its Japanese namesake; however, it may become similarly classic if its description of the enemy stands the test of time.

The Five Rings model describes the enemy as a set of concentric rings, ranging from the national leadership at the center outward to the fielded forces. The premise behind the theory—which, as mentioned, became the basis for the very successful air campaign against Iraq in Operation Desert Storm—is that attacking fielded forces diverts effort away from the main objective, which is to convince the national leadership to stop their hostilities.[11] Within each ring, we may identify critical nodes to attack;[12] the five rings model advocates simultaneous attacks on multiple targets in each ring to disrupt enemy operations as quickly and completely as possible. This notion is quite potent, even if it was not entirely original with Colonel Warden; in his near-classic text *Strategy*, B.H. Liddell-Hart also advocated multiple simultaneous attacks to confuse and confound the enemy.[13]

The Five Rings model could have an interesting alternate use. In addition to examining the subject state's organization to pick out critical nodes to attack, a fighting (and especially an occupying) force should be able use a similar model to find the critical nodes to *rebuild* after hostilities end. We might think of multiple models—an economic Five Rings, an informational Five Rings, and so forth—each with critical nodes. Whether our goal is to rebuild infrastructure for our own use, or to help the indigenous population recover from the war (and perhaps strengthen the moral component of our force, in the eyes of noncombatants), the Five Rings should help us prioritize reconstruction efforts by selecting reconstruction "targets" that are likely to produce the biggest gains. In some cases, the priority of rebuilding may be obvious, and no such analysis would be required; but where rebuilding needs outnumber the resources available, a little time spent on analysis could mean a lot of time saved and goodwill developed in the rebuilding effort. Thus the model may be used both for strategic attack and strategic rebuilding based on nodal analysis—and if the goal of strategic attack is paralysis, derived from the Greek for "by loosening," we might call the goal of reconstruction "paradenosis," since "deno" means "to fasten."

Earlier discussions presented how some of our military ideas and theories have changed over time (e.g., the addition of the operational level to the levels of war) and how they face challenges from time to time (e.g., the question of whether fog and friction continue to apply in the modern world). The Five Rings theory is still new enough that it may not reach its final form for some time. One possible weakness of the model as it was originally presented is that it does not indicate how the rings intersect and interconnect; such relationships between the nodes may be important to determining enemy weaknesses, in which case a system flowchart or other method of visualizing and categorizing the connections between the rings may be a useful addition.[14]

Even without that addition, however, the model still qualifies as a theory: it is a conceptual tool that describes enemy conditions and indicates ways in which the enemy system may be damaged. The Five Rings model is the most powerful of the military ideas we have explored thus far, and has been used to good effect in planning successful operations. In addition, it can be applied to any number of historical examples to discover why attacks in certain places did or did not produce the desired effects. A thorough study of the model may go far to establish its power as an explanatory and predictive tool.

False Theories. Other ideas and maxims masquerade as theories. The idea that anything in war—or, for that matter, in life—is "perfectly knowable and predictable" is one such fantasy. In advancing his thoughts on the Five Rings model, Colonel Warden boldly asserted that ". . . we are struck by the fact that the physical side of the enemy is, in theory, perfectly knowable and predictable."[15] This statement is hard to digest. Another way of writing the phrase is, ". . . the fact that, in theory, the physical side of the enemy is perfectly knowable and predictable." The construct, "the fact . . . in theory," is particularly troublesome.

At first glance Colonel Warden appears to say "we are struck by the fact that the physical side of the enemy is . . . perfectly knowable," which clearly is not a fact at all but rather a conjecture—and a poor one, at that. Then it appears Colonel Warden may be saying that "we are struck by the fact" that a theory exists; furthermore, the striking element of the theory is that the condition of the enemy should be perfectly knowable and predictable. Simplified further, this becomes "in theory, the physical side of the enemy is perfectly knowable and predictable." This is a fantastic notion, almost identical to the idea that we might ever "achieve total situational awareness"[16]—using the word "fantastic" deliberately, as in, derived from fantasy.

This bizarre notion relates to what Jomini called "military statistics," about which he wrote, "We have but vague and superficial statements." In discussing "means of gaining information," Jomini advised using several complementary means, "for, no matter how imperfect and contradictory they may be, the truth may often be sifted from them," and cautioned: "Perfect reliance should be placed on none of these means."[17] One might argue that modern observation techniques make Jomini's caution invalid, but would anyone claim those techniques to be infallible, or the people who employ them to be incapable of error? As just one example of the fallibility of modern techniques, consider that

> the "increase" in [chemical weapon (CW)] transshipment activity [in Iraq] that analysts saw starting in March 2002 may have been due, at least in part, to an increased volume of imagery *collected* by U.S. satellites rather than to any increased activity by the Iraqis. To only somewhat oversimplify the matter, it wasn't that the Iraqis were using Samarra trucks more often in 2002—it was that in 2002 the United States was taking more pictures of places where the Samarra trucks were being used. And this failure to distinguish between *actual* increased activity at suspect CW sites and the *appearance* of increased activity due to increased imaging likely contributed to the mistaken assessment that Iraq was ramping up CW production in 2002.[18]

This issue of quasi-theories recalls the observation that "Every theory is correct—in its own world."[19] Colonel Warden's idea, that *anything* is "perfectly knowable and predictable," let alone conditions of the enemy which we cannot observe directly and instantaneously, can be true only in a world vastly different from ours—and in that world, it would seem to make war almost impossible, as Martin van Creveld observed:

> . . . should information ever become perfect—should a complete mathematical model of the battlefield become available—then that model itself will already spell the end of the fighting. As in any game, where the outcome of a war can be calculated in advance, fighting does not make sense. . . . Such a situation permits armed conflict to be replaced by a computer [and] full-scale computerized warfare is still far away. . . .[20]

With respect to the reality we encounter daily, however, Colonel Warden's notion is nonsense. Modern physics tells us, for instance, that the quantum mechanical state of our universe has uncertainties built in from the outset. Therefore, the search for "perfect" knowledge of an enemy seems the search for a martial Schrodinger's cat.[21]

Another false theory is encapsulated in the cryptic sentence from one of the histories of the Principles of War: "In war correct methods are those that lead to victory."[22] Because this statement allows for nefarious, ruthless, or immoral means to achieve victory, the idea may be anathema to warriors for whom honor is an overriding virtue.[23] In addition, this is more a statement of belief than theory. It requires that every determination of a "correct method" be made after-the-fact, without benefit of analysis: If we won, we must have used the correct method, instead of, If we discover and use the correct method, we will win. Such reasoning is precisely the opposite of what theory is meant to do.

It may have occurred to you during some of this discussion that our thinking about war is often based on the enemy we face. Colonel Warden's Five Rings construct, for example, is first of all a representation of the enemy itself and second a guide to the type of forces needed to confront the enemy.

Focusing on the enemy is not of itself a bad thing; recall Sun Tzu's well-known admonition about knowing the enemy as well as one's self. Building an entire strategic approach on one enemy or one aspect of the enemy may be self-limiting, however, as in the case of the Mutually Assured Destruction approach to nuclear standoff during the Cold War. In a way, that limited focus on the destructiveness of nuclear weapons was beneficial; had our planners and decision-makers focused on their utility instead of their destructive power, the likelihood of their use may have increased. The single factor of the enemy's destructive potential did not afford a way out of that conflict, however. It was not until President Ronald Reagan pressed for including economic factors in the overall anti-Soviet strategy that the stalemate was finally broken.[24]

The "hearts and minds" campaign in the Vietnam War was another enemy-based theory of sorts, one that presupposed the possibility of converting or winning over noncombatants instead of directly defeating the enemy. This may be where Colonel John Boyd began considering the moral forces involved in war in relation to the mental and physical forces, and it is easier to turn the hearts and minds of an enemy against us than it is to bring them alongside us. Thus, when some characterized the approach to action in Iraq in this way—suggesting that the aim should be win the Iraqi peoples' hearts and minds—it was not the most promising sign.[25]

It is certainly easier to assess the outcome of an actual battle or maneuver than a campaign aimed at perceptions and attitudes, especially in a country with a diverse population; therefore, such a nebulous idea as "hearts and minds" does not have the makings of a good theory. In much the same way, if the strategic aim is larger than simply defeating the enemy's military, building the campaign around an enemy-based theory may be insufficient to the task: It is easier to plan and execute a military campaign than to orchestrate a complete economic and diplomatic overhaul.

In some cases our military acquisitions themselves are enemy-based, which can be problematic when the enemy against which we prepared either dematerializes (e.g., the Soviet Union) or turns out not to be the next enemy against which we have to fight. Every Quadrennial Defense Review undertaken by the US military wrestles with the best ways to provide combat capabilities, which are always best understood in relation to the range of possible enemies. But the study of acquisitions and how best to prepare for the unknown future—especially within the strictures of Federal laws and regulations—is far beyond the scope of this text.

Recall from the definition we posited in Chapter 2 that war requires both an enemy and an aim. Accepting "the enemy" as a given may be enough, at least to start. To focus too much on the enemy holds two dangers: one, that the aim may not be given enough attention and be missed or not fully realized; two, that we may not be prepared for the appearance of new enemies. Enemy-based theories, then, are reactive and defensive instead of proactive and offensive, but they are still better than no theory at all.

Most of these "quasi-theories" fall far short of what we expect of theory, in the sense of Theory of Knowledge, as a way of categorizing experience. A good theory will be useful in explaining the past or predicting the outcomes of our actions, but these would-be theories are inadequate to explain what happened before, and cannot help us predict with much confidence what may happen in the future. Let us hope other military theories better meet those needs.

18

CONSIDERING THE "PRINCIPLES OF WAR"

If we are right when we claim that daily life requires constant theorizing about situations and events, even when that theorizing goes unrecognized, then surely the critical and chaotic enterprise of war requires continued and concentrated theorizing to ensure our sustained readiness for it and improve our chances for victory. We have already defended this view and this practice—as well we might—from an epistemological as well as an historical perspective.

Furthermore, as we go through the process of formulating theories and then testing those theories against reality, we should not be surprised sometimes to discover and have to deal with weaknesses in some of our theories. When we start with existing theories, as we have done so far with the levels and media of war, we may try to reformulate them to overcome the weaknesses and enable us to make better predictions and gain a more thorough understanding. Our next step is to apply that sort of process to the theory formed by the well-known "Principles of War": Objective, Offensive, Mass, Maneuver, Economy of Force, Unity of Command, Security, Surprise, and Simplicity.

Considering the discussion of the nature of theory, and specifically the nature of military theories, the Principles of War may be thought of as either a single combined theory or a collection of separate theories about the conduct of warfare. Some may argue to the contrary, that they do not form a theory at all, and make a convincing case; Joint doctrine, for example, does not afford them that status. In fact, Joint doctrine expands the traditional principles to produce a set of "principles of joint operations" that, according to the 2007 version, "combine the nine historical principles of war (present in joint doctrine since its inception) with three additional principles born out of experience across the range of military operations": viz., Restraint, Perseverance, and Legitimacy.[1] But the 2017 *Doctrine for the Armed Forces of the United States* "capstone" document mentions the Principles of War only briefly, and relegates their explanation to *Joint Publication 3-0, Joint Operations*—where they are primarily treated in an appendix.

Beyond Joint doctrine, service level doctrine is quite uneven in its treatment of the Principles of War.

Army Doctrine Publication 1-01, Doctrine Primer, refers to the larger group of joint operations principles as "time-tested general characteristics of successful operations that serve as guides for the conduct of future operations."[2] Even so, the principles do not form a prominent part of Army doctrine. *ADP 1-01* does not go into the principles in any detail, and the 2017 version of Army *Field Manual* 3-0 does not mention them at all. Earlier iterations of FM 3-0, in contrast, listed the principles of war as one of "several sources of knowledge and relevant information" that help commanders visualize the desired end state of a mission, and stated that they "represent the most important nonphysical factors that affect the conduct of operations at the strategic, operational, and tactical levels."[3] Despite calling them "the most important . . . factors," the 2008 version of *FM* 3-0 placed the principles in a minor role relative to, for instance, the elements of combat power—movement and maneuver (itself reflecting one of the classical principles), fires, intelligence, leadership, protection (somewhat analogous to the principle of security), mission command (formerly "command and control"), and sustainment. In contrast, the 2001 version treated the elements of combat power and the principles of war almost equally.[4]

Even though their influence in Army doctrine has waxed and waned, it is important to note that *ADP 1-01* does state that the "greatest value" of the principles "lies in educating the military professional. Applied to the study of past operations, the principles of joint operations are powerful tools that can assist commanders in analyzing pending operations,"[5] which is precisely what we expect of sound theorizing: that it will distill experience into concepts that we can use to predict future outcomes. If that is truly the case, it certainly seems appropriate to consider the principles as comprising a theory.

In Air Force doctrine as well, the Principles of War have not always been treated consistently. The old Air Force Manual 1-1, for instance, seemed to dispute the idea that the principles form a theory: "The so-called principles of war merely represent generally accepted 'truths' that have proven effective for commanders employing forces in combat."[6] The qualitative "merely" gives an indication of the relative merit ascribed to the principles, but the passage continued, "They offer a conceptual framework within which to evaluate possible actions," which again is *precisely* the role of a theory. Indeed, Air Force doctrine's description of the principles as "valuable guides to evaluate potential courses of action" aligns quite well with the role of theory as we have portrayed it thus far: a tool to explain the past and predict the future.[7] However, the fact that the Air Force added another principle beyond those already appended by the Joint Staff—viz., Unity of Effort—would seem to bolster the case that in some respects at least the classical Principles of War do not form a very *complete* theory.

Not surprisingly, the Marines' doctrine contains some clear thinking on the subject. Though they are again the subject of an appendix, *Marine Corps Operations* introduces the classical principles with the mnemonic "MOOSEMUSS" and calls the twelve joint operations principles

> useful aids to a commander as he considers how to accomplish his mission. They assist the commander in organizing his thinking about his mission, the enemy, the battlespace, and his forces. They should not be considered as prescriptive steps or actions that must be accomplished, but as tools to plan, execute, and assess operations.[8]

Does that not sound very much like what we expect a good theory to do?

The Principles of War do not describe *what* happens in war so much as they attempt to describe *how* things happen in war. They do not rise to the level of if-then propositions, since all of the "ifs" cannot be fathomed ahead of time, but they are an attempt—albeit disordered—to describe as well as predict the outcomes of military actions. As such, they fit quite well the description of theory we previously presented.

Just as it may be debated whether the principles form a theory, the principles themselves—i.e., the nine classical principles—are not as universally accepted as one might imagine. While Jomini claimed, "Every educated military man will be impressed by the truths educed" in his defense of his central principle (see "Mass" in Chapter 19),[9] Clausewitz disparaged them, asking: ". . . what is the use of such feeble maxims? They belong to that wisdom which for want of anything better scribblers of systems and compendia resort to when they run out of ideas."[10]

Even as he decried the result, Clausewitz lauded the effort: "Efforts were . . . made to equip the conduct of war with principles, rules, or even systems. This did present a positive goal, but people failed to take adequate account of the endless complexities involved."[11] Some of those complexities will be discussed in succeeding chapters. And while eminent historian Russell Weigley stated, "the principles are so straightforwardly simple and general that they can lead astray almost as easily as they assist,"[12] even recent theorists fall back on them again and again.[13]

The principles are not as changeless as they seem, either. Military historian John I. Alger chronicled their development, showing how they grew beyond Jomini's inception of his central principle.[14] Alger's brief chronology of their more recent changes in the US runs as follows: the Army's nine principles of war in the 1920s became ten principles of offensive combat in 1931; by 1936, they had been reduced to seven principles of strategy and by 1939 to six principles of war; then the nine principles of war, the same list we have today, appeared in 1949.[15] In addition to those nine "traditional" principles, in the 1990s Joint Doctrine included another set of principles to cover "military operations other than war," or MOOTW, a term that has since been dropped from Joint usage.[16] At one time two new principles, flexibility and logistics, were proposed to be added to the canon, but as discussed current Joint Doctrine only includes three additional principles—restraint, perseverance, and legitimacy—in the total "principles of joint operations" construct.[17] Thus it is imprecise to say, as USAF Colonel John Warden did in *The Air Campaign*, that the principles "have not changed since Miltiades repulsed the Persians on the Plains of Marathon."[18] (Furthermore, considering that Miltiades defeated a Persian army three times larger than his own, his victory is hardly a ringing endorsement for the application of mass.)

But despite challenges and changes, most of the principles are accepted, as Colonel Warden says, "perhaps on a visceral level."[19] A fair question is whether the fabled "gut check" is the best basis for accepting a theory.

Recall that a good theory must adequately explain the past as well as predict the future. To evaluate the Principles of War as a theory, then, we may examine each one in light of examples from military history. Dedicated proponents of the principles should assert that any such endeavor cannot fail to uphold them, as "their existence is detected by the study of the past, which reveals them in successes and in failures, the same from age to age."[20] However, Theory of Knowledge as we have discussed it holds that any failure of a theory—or in this case a principle —to explain or predict an outcome requires that we either reformulate or reject it, since a theory should not be based on only those historical examples specially selected to support it: "A careful study of history will illustrate that principles are not immutable rules which the commander is forbidden to violate. Nor should a theory be based on historical examples arbitrarily selected to support an unfounded preconception."[21]

We should pause for a moment to emphasize this point. History, with respect to proving or disproving a military hypothesis, "is a seductive mistress," because "it is all too easy to . . . find, in history, evidence (examples) that 'prove' almost anything we wish to prove."[22] On the other hand, the search for historical analogies can lead down unexpected paths.

> Military historians are exasperating fellows; they profess to help the decision maker, the activist military commander, to see more deeply into his problem. They are exasperating because, instead of simplifying the commander's problem, they only show him how much more difficult it is than it appeared at first.[23]

(You may well agree, when all is written and read, that this very text is just as exasperating for much the same reason.)

If we must have a care with respect to referencing historical examples, under what circumstances might a generally-accepted principle of war be invalid? Defenders of the principles would hold that "the battles of the past succeeded or failed according as they were fought in conformity with the principles of war,"[24] so almost any evidence to the contrary should be enough to call a principle into question. If we find instances in which a principle fails to explain the outcome of a battle or campaign—whether the principle was applied and the battle lost, or the battle won even though the principle was not applied—then the principle may require refinement in light of that failure.

Before we examine each principle on its own merits, we must recognize the weakness of considering historical events from the perspective of a single viewpoint.[25] Seldom does one factor alone determine the outcome of a complex web of events. Indeed, fog and friction, because they complicate the warfighting situation, effectively deny causation to any single principle of war and require that interactions between the principles be considered. Even *Air Force Basic Doctrine* warns, "No one principle should be considered without due consideration of the others."[26] Realistically, however, the interactions between the principles are too numerous and complex to consider in a short review. We will return to the interactions in Chapter 22, but meanwhile let us continue this process while remaining mindful of its limitations.

19

THE NINE CLASSICAL PRINCIPLES OF WAR

How are we to evaluate the Principles of War as a theory? To begin with, by taking each principle individually and considering it in terms of how well it explains the outcomes of past engagements.

Objective. Joint Publication 3-0 notes that an objective must be specified in order to "direct every military operation toward a clearly defined, decisive, and achievable goal."[1] While this "frequently involves the destruction of the enemy's capabilities and their will to fight," current doctrine provides for broader goals "not involving this destruction." JP 3-0 adds, "Objectives must directly, quickly, and economically contribute to the purpose of the operation. Each operation must contribute to attaining strategic objectives."

Since an overall objective is necessary to war itself, as stipulated in Chapter 8, having a definite objective seems appropriate and necessary to successful military operations. Certainly Seneca clearly stated the objective in the Roman Senate when he said, "Carthage must be destroyed." In 1991, the coalition forces in Operation Desert Storm had a clear objective as outlined by UN resolution, and fulfilled it by driving the Iraqi army out of Kuwait; that objective probably was not the best possible one, since it left Saddam Hussein in power and turned a 6-month campaign (including Operation Desert Shield) into a 12-year undeclared war, but it had the benefit of being well defined. The Terror War and its continuing analogs demonstrate the difficulty of formulating concrete objectives when the enemy is both bellicose and amorphous; still, without some objective in mind no military power can be focused.

Familiar—and negative—aspects of the Vietnam experience also point out the need for a defined objective. Earlier we mentioned the effort to win Vietnamese "hearts and minds," but that was not a clearly stated objective, nor did it contribute directly to military aims; likewise, the "hearts and minds"-like program in Iraq was aimless rather than "a clearly defined, decisive, and achievable goal." One obvious difference was that military objectives in Vietnam were determined as often by political expedience as by military necessity. Another difference was that a new civil government had to be established in Iraq, which required close interaction with the populace and civil authorities. That effort suffered, though, because in the rush to Baghdad we did not pay enough heed to the admonition that all politics is local.

Does history present any instances where competent commanders, going after clear objectives, have failed? We might list candidates—Germany's invasion of the Soviet Union in the early days of Operation Barbarossa, for example—but their failures to achieve the objective cannot be pinned on the existence of the objective. In the case of Barbarossa the objective changed, but changing the objective is the commander's prerogative, as shown by General George S. Patton's famous statement: "I've been studying the art of war for forty-odd years. When a surgeon decides in the course of an operation to change its objective . . . he is not making a snap decision but one based on knowledge, experience and training. So am I."[2] Sometimes the objective must change because the enemy's center of gravity changes (or becomes more discernable than at the beginning), and sometimes the objective changes for less clear reasons. JP 3-0 notes that "changes to the military objectives may occur because national and military leaders gain a better understanding of the situation, or they may occur because the situation itself changes."[3] Again, however, we may reiterate that failure to achieve an objective does not depend on the objective itself.

In any example, myriad factors—fog and friction—combine to either expedite or impede success, and very few of them owe their influence to the objective being sought. Thus having a clear objective is important, but may not be enough; so, "objective" falls short as a theory. It does not adequately explain the success or failure of past encounters, nor does it satisfy as a predictor of success.

If "objective" as a concept does not meet the criteria for a sound theory of warfighting, it still cannot be discounted. Indeed, as pointed out in Chapter 8 it seems self-evident: We do, after all, go to war to achieve some stated end. Having an aim is essential, but the aim, and the accompanying prediction of success, should be *based on* theory rather than *being* a theory unto itself.

In setting objectives, Liddell Hart warned, ". . . to ensure reaching an objective one should have alternative objectives. . . . For if the enemy is certain as to your point of aim he has the best possible chance of guarding himself—and blunting your weapon."[4] Every objective, therefore, should be selected based on a sound evaluation of the situation that supports the prediction that the effort will succeed (since such a prediction is inherent in the simplest of theories), but with the understanding that the prediction may be wrong.[5] The object of subsequent action, of course, is to make the prediction come true; but that depends on factors beyond the objective itself.

Offensive. "Seize, retain, and exploit the initiative" as "the most effective and decisive way to achieve a clearly defined objective."[6] Here is another principle that seems essential: we cannot, after all, defend our way toward the goal. Two famous examples of successful offensives are Sherman's march to the Atlantic during the US Civil War and the Allied campaign in North Africa in 1942. These offensives, and others like them, echo the words of Sun Tzu: "Invincibility lies in the defense; the possibility of victory in the attack."[7]

For thousands of years, war required the offensive and the close-in action of soldier versus soldier. Developments in weaponry, however, made such offensive action a much more dangerous proposition. The result is a principle that may hold true at higher levels of war but is less reliable in many tactical situations.

For example, consider that the French tried to take the offensive—to seize the initiative, to act rather than react—when they charged the

English at Crecy and Agincourt and were cut down in droves by arrows from Welsh longbows. Those armies were similarly trained and equipped, except for that single weapon at the English army's disposal. In a somewhat different vein, tribesmen used to a traditional mode of combat often chose to attack their imperialist enemies directly, only to be slaughtered by machine guns:

> Archetypical was the 1898 Battle of Omdurman in the Sudan, where the disciples of the Mahdi, the fabled dervishes, repeatedly hurled themselves against British lines, only to be repulsed each time by six Maxim guns firing six hundred shots per minute. . . . An even more lurid example . . . took place in Tanzania during the early 1890s . . . a German officer-surgeon and his assistant dragged two machine guns and plenty of ammunition into a mud hut with a clear field of fire, and from there killed around a thousand native combatants. Two men, two guns, a thousand dead—if ever there was a device capable of giving vent to man's predatory instincts, it was the machine gun.[8]

It could be argued that the French and native armies did not apply the principle of the offensive correctly, but that argument is less a critique of those armies than a defense of the principle itself against the realities of modern warfare. As John Keegan pointed out, "The appearance of the machine-gun, therefore, had not so much *disciplined* the act of killing—which was what seventeenth-century drill had done—as *mechanized* or *industrialized* it."[9] As with all such revolutionary advancements in the art of killing, it took time to assimilate and understand what it meant for warfare as a whole.

It may be said that the offensive is as necessary to war as the objective. Without an offensive, what chance victory? In today's environment, however, the offensive may take different forms. The drive to Baghdad may be as likely as the village-by-village, mountain-by-mountain offensive in Afghanistan. And the "offensive" taken by the bellicosi is more likely to be guerrilla- or saboteur-oriented than an open attack on a military target—although Al Qaeda mixed an attack on a strictly military target, the USS *Cole*, into their attacks on civilian (the World Trade Center) and government (US embassies) targets, though few recognized the *Cole* attack as an act of war.[10]

Mass. "Concentrate the effects of combat power at the most advantageous place and time to achieve decisive results," according to *Joint Publication* 3-0.[11] Never mind the semantic difficulty of concentrating an *effect* rather than concentrating combat power itself: effects have causes, but the definition skips over the (concentrated?) causes that would produce the concentrated effects. It simply adds that mass is achieved when combat capabilities "are integrated and synchronized where they will have a decisive effect in a short period of time."[12]

Joint Publication 3-0 includes a definition of "combat power" in a glossary, rather than with the definition of mass (or with any of the other principles that employ it). The same definition is in the *DOD Dictionary of Military and Associated Terms*: "The total means of destructive and/or disruptive force that a military unit/formation can apply against the opponent at a given time."[13] So the definition of mass, using that of combat power, may be taken as, "concentrate the effects of the total force which can be applied at a given time, at the most advantageous place and time to achieve decisive results." The critical element in this definition appears to be not combat power itself, but rather the temporal and spatial application of that power. We will discuss that in a later chapter. In addition, the total force which can be applied may not be the total force available; this is the essence of the next principle, economy of force, to which JP 3-0 alludes when it states, "Massing effects . . . rather than concentrating forces, can enable even numerically inferior forces to produce decisive results."[14] One wonders how any principle so defined can be completely understood, let alone correctly applied.

Although the principle of mass is as ancient as warfare itself—the defenders at Thermopylae, while their weakness was revealed by a traitor, were eventually overwhelmed by sheer numbers—its modern Western expression began with the rise of the common soldier during the Napoleonic era. In the Twentieth Century, the principle was applied by the Soviet Union against Nazi Germany at the expense of millions of lives, and later by the Chinese when they sent 300,000 troops to confront the US Army in Korea in what was only a short-lived victory. It is no wonder that mass has long been a favorite of martial thinkers, especially in its incarnation as the principle of "concentration."[15]

Mass was Jomini's "one great principle underlying all the operations of war . . . which must be followed in all good combinations . . .", and he stressed action against "fractions of the hostile army with the bulk of one's forces."[16] Despite the fact that Jomini wrote of Napoleonic columns, the elegant simplicity of such statements quickly gained adherents, and application of the principle of mass to naval and air operations grew as prevalent as to ground warfare.

In naval theory, Mahan stressed the importance of massing resources against a portion of the enemy fleet: "divide the opposing force by penetrating through an interval found, or made, in it, and then . . . concentrate upon that one of the fractions which can be least easily helped by the other."[17] Yet Mahan argued against procuring huge and powerful dreadnoughts, which might have been considered the next logical step in building and wielding maritime mass.[18] While Mahan would lose that argument, his misgivings would eventually be vindicated by the experience of World War Two.[19] Indeed, if "concentration" was really the guiding principle of naval warfare, the progression from battleships to dreadnoughts should have led to the "floating fortresses" mentioned in George Orwell's *1984*—especially if concentration was superordinate to mobility.

Mass is inherent in most airpower theories from the earliest strategic bombing studies to modern—especially nuclear—applications.[20] One question that arises from a discussion of strategic bombing is, what makes some bombing "strategic"? In general, the answer is the targets that are being bombed, although in Chapter 12 we mentioned the weapons-driven outlook that came into being with the development of nuclear weapons. Whither, then, "tactical" or even "operational" bombing? When discussing battlefield applications of airpower, we seem to prefer more descriptive if not more exciting terminology such as "air interdiction" or "close air support." What, then, of Liddell-Hart's "grand strategic" bombing? We pose these largely rhetorical questions to point out that even the naming conventions we use are attempts to explain and/or understand the world around us, and so constitute rudimentary theories.

The growing influence of "smart" weapons and stealth platforms might at first seem to invalidate the principle of mass, except that the traditional principle is not "mass" as in a large force, but "concentration" as in a large force per unit area: one bomb dropped precisely on target can be more devastating than tons of bombs that miss.[21] Colonel Phillip

S. Meilinger, for example, argues that airpower—and especially the precision-guided munitions that airmen now employ—has simply changed the nature of the concept rather than changing the concept itself or rendering it obsolete; to illustrate the change, he introduces "density" as a preferred term.[22]

Regardless of whether one prefers the name "mass" or "concentration," the root of the principle lies in superior force; but despite many examples of success, this principle is not foolproof. Sun Tzu recognized this, warning, "In war, numbers alone confer no advantage. Do not advance relying on sheer military power," and pointing out that attacking larger forces requires extreme care and calculation.[23] Vegetius likewise stated, "In every battle it is not numbers and untaught bravery so much as skill and training that generally produce the victory."[24]

Clausewitz wrote, "to accept superiority of numbers as the one and only rule, and to reduce the secret of the art of war to the formula of numerical superiority *at a certain time in a certain place* was an oversimplification that would not have stood up for a moment against the realities of life."[25] Clausewitz seemed to aim his words directly at his contemporary Jomini, and he and Jomini sparred over their respective ideas. Jomini, for his part, resorted to personal attack, e.g., calling Clausewitz's logic "frequently defective."[26] However, Jomini almost refuted the mass principle himself, using examples where smaller armies defeated larger ones by attacking first one flank, then the next, and so on.[27]

To address this issue by means of a simple thought experiment, if mass truly predicted success in battle or in war, and all potential combatants understood that, there would be no battles and no wars. (As mentioned in Chapter 17, Martin van Creveld discussed this in *The Transformation of War*.) Conflict would be a simple matter: compare numbers and capabilities and concede to the larger or more powerful force, then run home and do all you can to create more combat power. In the end, the most prolific society or the best consensus or alliance builders would always and automatically win. The logic seems solid enough, if the predicates themselves are firm. It is easy to make such statements, to rest on logic as if it were a sure foundation, except that human behavior is not so precisely logical (least of all war, which may be the most illogical activity of all). But in reality the predicates themselves are shaky because mass does not always predict success.

In addition to warnings from noted theorists, historical examples counter to this principle throw doubt on its validity as an axiom. A very famous ancient example is Alexander's victory over Darius at Issus, where the Macedonian prince led his outnumbered troops across the river into the heart of the Persian army. Another ancient example involves Hannibal's campaign in Italy: his army was reduced from 38,000 to 19,000 crossing the Alps, but they surprised and defeated Roman legions numbering around 150,000 in the battles of the River Trebia, Lake Thrasymene, and Cannae.[28] It is important to note that Hannibal and his army did not, despite these successes, win the war, an observation that returns to the heart of the issue of tactics and strategy.

Ancient examples do not always convince, however, especially since Hans Delbruck showed that many historical accounts—including, for example, the description of the battle of Cannae—are inflated.[29] More recent counter-examples, with perhaps more verifiable figures, come from Spanish experience in the Americas and Russian experience in the Crimea. Pizarro's capture of the Inca emperor Atahuallpa at Cajamarca in 1532 may not be much more convincing, because no pitched battles were involved: certainly Pizarro's 62 horsemen and 106 foot soldiers should not have been able to defeat Atahuallpa's 80,000 men, but they secured their victory thanks to Spanish guns and armor, aided by the epidemic smallpox that had divided the Inca empire and to which Pizzaro's troops fortunately were immune.[30] So we turn to still later examples. In the late 1700s, Russian armies, vastly outnumbered, soundly defeated the Turks on two separate occasions.[31] In contrast, in the Crimean War in the 1850s, "The mobilization of 1,742,297 officers and men (plus 787,197 irregulars and militia) proved inadequate to deal with a force of 300,000 French, British, Sardinian, and Turkish troops."[32] In addition, Liddell Hart includes in his classic text *Strategy* several examples of numerically inferior forces emerging triumphant.[33]

Applying the principle does not, then, guarantee victory. It may improve the odds, however. So, while "mass" does not carry the weight of a categorical imperative, its appeal to common military sense remains strong and its defense vigorous. Jomini asserted that "Every maxim relating to war will be good if it indicates the employment of the greatest portion of the means of action at the decisive moment and place,"[34] but that cannot be said to state an ultimate truth. Jomini also wrote,

> The application of these fundamental principles is also very simple. If you have one hundred battalions against an equal number of the enemy's, you may, by their mobility and by taking the initiative, bring eighty of them to the decisive point while employing the remaining twenty to observe and deceive half of the opposing army. You will thus have eighty battalions against fifty at the point where the important contest is to take place.[35]

Jomini neglects to mention that, in this hypothetical case, the enemy also has fifty battalions against your twenty, should they take the initiative and should you choose a "point" that is less than decisive. The "decisive point" may not always be the same for us as for our enemy.

Based on critical thought and the plethora of contrary examples, then, mass fails as a theory; however, we may at least accept superior mass as a factor that increases the probability of success. From this we may derive any number of lessons, such as: do not wait too long to build or acquire overwhelming force; do not surrender too early when faced with apparently overwhelming force; and more is, in the end, better than less (as with many things; e.g., money). Thus we recognize the potential advantage of mass while avoiding the danger described by Liddell Hart of "the tendency of textbooks to treat war as mainly a matter of concentrating superior force."[36]

As mentioned earlier, why a small force may defeat a larger force may lie in the next principle.

Economy of Force. "Expend minimum essential combat power on secondary efforts in order to allocate the maximum possible combat power on primary efforts" by allocating combat power to "limited attacks, defense, delays, deception, or even retrograde operations to achieve mass elsewhere at the decisive point and time."[37]

This is a curious principle. As with the previous principle, it includes "combat power," but this time directed at efforts other than the prime objective. The pursuit of secondary objectives, even with the minimum of combat power spared to them, implies splitting forces intentionally, which can backfire. The rationale for allocating minimum forces to such efforts is twofold: first, to preserve the main force for the main objective, and second, to reduce the harm should the secondary effort fail. Still, it seems safe to ask why this principle and the principle of mass are even separate. As USAF Colonel Edward Mann wrote, "One cannot completely remove military effort from all areas save one, because doing so would create unacceptable vulnerabilities in the vacated areas. Rather, one must strike a balance. . . ."[38]

Economy of force and mass, then, are implicitly linked. Is that link unbreakable? Perhaps; at least it seems difficult if not impossible to have one without the other. The 2000 edition of *Joint Publication 1* gave mass and economy of force as examples of principles that "paired juxtapositions that address opposite aspects of warfare."[39] Mass, especially as demystified above, could perhaps be a definition supporting a more comprehensive version of economy of force. Or it could be a corollary principle.

Commanders can succeed when using available forces judiciously, as shown in numerous examples of victories by numerically disadvantaged forces against superior foes, from Miltiades and Themistocles to those cited above. Certainly, the Israeli army demonstrated remarkable economy of force in the various Arab-Israeli conflicts, often defeating numerically superior but uncoordinated foes. This principle was also well applied during the Italian campaign of World War II, as the German army retreated up the peninsula. This last example might lead us to believe that economy of force applies best to defensive and siege campaigns; however, because many defenders capitulate under siege, it cannot be said to be all-inclusive. Indeed, Sun Tzu's admonition—to use the enemy's attempt to prepare in many places against him by selecting a few places to attack—seems to cut to the heart of economy of force.[40]

"Economy of force" helps explain *why* the principle of mass (or concentration) is not the perfect predictor that Jomini intended. It is not exactly useful as a guide for action, though as an explanation of the past it works fairly well. If we held to it too closely, the pursuit of economy of force would seem to tie the hands of battlefield and higher headquarters commanders: the quest for the ultimate economy may lead them to hesitate or avoid action altogether for fear of taking casualties. In contrast, the need for action should not be taken as license to waste forces. But since using one's forces judiciously does not guarantee those forces will win, and using forces carelessly does not predict defeat, economy of force fails the requirement for a successful predictive theory.

We do our preparedness a disservice when we overemphasize economy of force. If producing surplus is essential to sustaining a working economy, much less economic growth, over the long term,[41] then producing a military surplus may be just as essential to sustaining a working military. In this era of high-tech, high-cost, long-lead-time weapons, what happens when the inventory runs out? Perhaps the aim should be surplus in preparedness, economy in use.

Even so, economy of force is clearly evident in much of conventional (and some unconventional) wisdom. It makes sense not to trim the roses with a chain saw and not to nuke Baghdad to get rid of the dictator. In other words, use no more force than needed but be quite sure you know what is needed, as Miyamoto Musashi noted in 1645: "When you sacrifice your life, you must make fullest use of your weaponry. It is false not to do so, and to die with a weapon yet undrawn."[42] And remember the wisdom of the playground: "Close only counts with horseshoes, hand grenades, and nuclear weapons."

Economy of force is also fundamental to terrorists and bellicosi, where a lone bomber or small group of hijackers can create disproportionate effects. Discounting their rare strikes against military targets (e.g., the Marine barracks in Beirut, Air Force housing at Khobar Towers, the USS *Cole*), the bellicosi achieve such effects because their attacks are usually against soft targets and noncombatants. In their view of the world, they may consider that there are no noncombatants, that women and children are legitimate targets, but this only shows how skewed their view of the world really is.

Maneuver. "Place the enemy in a position of disadvantage through the flexible application of combat power."[43] Again, with the inclusion of "combat power," the wording of this definition recalls the Jominian emphasis on mass, though not as definitively as his own words: ". . . the excellence of maneuvers will depend upon their conforming to the principle already insisted upon; that is to say, the great part of the force must be moved against one wing or the center, according to the position of the enemy's masses."[44]

As with other principles, the sheer simplicity of the idea is both appealing and limiting. Certainly, physical translation—motion—is absolutely necessary to military operations. An army must move to encounter its foe or approach its objective, a navy cannot remain in port, an air force unable to fly is useless, and spacecraft in orbit cannot help but move continually. But is the ability to maneuver well and successfully, enough upon which to predict victory? For example, the maneuverability of Greek vessels figured in their victory at Salamis, but that was not the only contributor.

Advances in weaponry prior to World War One made much Napoleonic-style maneuvering obsolete, as in the "race for the channel" that devolved into trench warfare. As pointed out eloquently by John Keegan: ". . . like a thunderclap, the French counter-stroke was launched, the first Battle of the Marne lost, the German spearheads sent trundling into reverse and the urgent footfalls of manoeuvre warfare drowned by the thud of spades digging the first trenches of the Western Front."[45] It is interesting to note that the Eastern front never stagnated as did the Western, yet neither were great wheeling maneuvers of troops employed.[46]

So this principle remains intact, largely because it is necessary, and we are cautioned by Sun Tzu: "Nothing is more difficult than the art of maneuver."[47] Like the other principles, we can acknowledge the necessity of maneuver, but as a theory it lacks predictive or explanatory power. Though we might take it to the extreme and say the complete inability to maneuver is likely to presage disaster, we cannot state categorically that those who maneuver better, win, and those who maneuver worse, lose.

With respect to the idea of modifying theories to fit experience, the modern era may require a new interpretation of this principle. Since the principle was formerly known as "movement," perhaps now "speed" might be the preferred term. This is not a new suggestion, however: John Alger pointed out that movement was equated with "rapidity" by British Major General and theorist of armored warfare J.F.C. Fuller;[48] and Sun Tzu wrote, "Speed is the essence of war."[49]

More recently, USAF Colonel Phillip Meilinger discussed the effects of speed and pointed out how airpower's "conquest of time" can be a deciding factor in many encounters.[50] Tu Yu, one of Sun Tzu's many commentators, wrote, "An attack may lack ingenuity, but it must be delivered with supernatural speed,"[51] which calls to mind General Patton's line about a good plan executed now with audacity. Today we rely on the "supernatural speed" of aircraft and a variety of guided missiles. Indeed, the possibility of conventionally-armed ballistic missiles, or other kinetic weapons dropped from orbit, screaming onto a target at many times the speed of sound virtually defines the term.[52] One of Sun Tzu's recent champions, Mark McNeilly, devotes an entire chapter of his *Sun Tzu and the Art of Modern Warfare* to what he describes as the principle of "moving swiftly to overcome resistance." He notes that "Speed in warfare provides four advantages: it is a substitute for resources, it shocks and surprises the enemy, it is critical to exploiting weaknesses and opportunities, and it builds momentum."[53] The US Army's old *Field Manual 1* stated that speedy action "dislocates the enemy force . . . preempts threats to security. . . . reduces suffering and loss of life . . . . gives Army forces their expeditionary quality . . . . [and] allows Army forces to keep the initiative."[54] And the underlying idea of USAF Colonel John Boyd's OODA Loop was that the pilot able to observe, orient, decide, and act faster would be more likely to win.

Unity of Command. "Ensure unity of effort under one responsible commander for every objective."[55] Like most principles, this seems self-evident—yet we fiddle with command structures and relationships almost constantly to keep it in place.[56] Certainly lack of unity of command hurt the Allied campaign in North Africa, especially with respect to the use of airpower in support of ground forces. The coherence of action in Operation Overlord can chiefly be attributed to its being led by a single individual, even though coalition political wrangling played a part in selecting that leader. The same might be said of Desert Shield/Desert Storm, an example of success despite General Schwarzkopf's carefully considered decision to violate the principle by allowing the Arab forces to be commanded by a Saudi general.[57]

As with all the other principles, the questions governing this analysis are whether adhering to it promises victory, or violating it spells defeat. The answer to both seems to be no. A strongly centralized command structure did not give Hitler any advantages during World War Two, and indeed sometimes hampered the efforts of his forward commanders; in Chapter 17, we discussed whether modern, highly-interconnected command structures may be setting themselves up for the same type of paralysis. In contrast, the Viet Cong successfully waged war against the US using a decentralized, networked command structure.

With respect to command structure, particularly complex structures have developed over the years in the US under the watchful eye of Congress, directed by legislation such as the Goldwater-Nichols Act ("Title 10"). In the totality of these structures, unity of command is stretched into almost unrecognizable shapes, but the underlying purpose is not to confuse but to ensure that lines of authority are clear for our various forces. The unwieldy constructs of administrative control (ADCON), tactical control (TACON), operational control (OPCON), etc., came into being precisely because our forces are enormously complex; exercising combatant command (COCOM) over the modern, Joint military is no easy feat.[58] Indeed, our increasingly information-based world is likely to require more networked/matrixed leadership structures rather than traditional hierarchical approaches.[59]

Security. "Prevent the enemy from acquiring unexpected advantage," which "enhances freedom of action by reducing friendly vulnerability to hostile acts, influence, or surprise."[60] This should apply to all forms of security (physical, operational, communications, and so forth) because all can be compromised.

Famous historical examples abound of lapsed or broken security—e.g., the Polish and British experience breaking the German Enigma code, the US advantage at Midway thanks to a broken Japanese code.[61] We also find famous examples of successful security, such as that of the US island-hopping operation thanks to the secure communications provided by Navajo code-talkers. Other examples do not involve cryptography: the Japanese knew a great deal about the Sunday morning routines at Pearl Harbor, though their ignorance of the location of US Navy aircraft carriers counted greatly in the long run. And we can find examples where security compromises endangered but did not doom operations; e.g., Enigma decrypts warned the British about the assault on Crete, but the intelligence was incomplete and Crete fell to the German forces.[62]

Unfortunately, the pre-attack preparations of Al Qaeda provide an example of successful security. Good security is essential for terrorist actions, bellicosi attacks, and all "revolutionary" conspiracies; not perfect security, but good enough to get their missions accomplished. The hearings and accusations regarding the US intelligence "failures" were quite interesting, not because no failures occurred but because it was apparent that it would not do to credit the enemy with a success. Whether further investigations would have helped improve US intelligence or preparations for future attacks is debatable; how effective were the Pearl Harbor investigations at anything but assigning blame for ill preparedness?

It is a given that the enemy will try to ascertain our plans and abilities; at the same time, we try to determine theirs. This aspect of security is one of the few points of agreement between Clausewitz and Jomini, both of whom pointed out that the enemy's situation cannot be known with perfect reliability (as opposed to Colonel Warden's assertion that it can, discussed in Chapter 17).[63]

Our determination of the enemy's plans includes another corollary: the use of espionage and various methods of intelligence gathering. Spying was particularly important to Sun Tzu, who devoted the final chapter of *The Art of War* to the subject: "What is called 'foreknowledge' cannot be elicited from spirits, nor from gods, nor by analogy with past events, nor from calculations. It must be obtained from men who know the enemy situation."[64] It is interesting to speculate what Master Sun would have written about the use of reconnaissance satellites and other means of intelligence gathering. Perhaps he might warn against over-reliance on them, and point out that the strengths of high-technology means are different from low-technology methods; likewise, the commission investigating the 11 September attacks stressed the need for improved human intelligence gathering in addition to our high-tech collection systems: "While imagery will be a valuable tool for the Community in developing a full picture of a target country's infrastructure and overt movements, without credible human or signals intelligence imagery is of limited utility with regard to [biological weapons] and [chemical weapons]."[65]

To ensure the enemy does not receive reliable information about our situation, a prime corollary to security is *deception*, the art of fooling the enemy: as Sun Tzu wrote, "All warfare is based on deception."[66] The use of deception is as old as warfare itself; consider, for example, Belisarius's insistence that his soldiers "spread out over the plain and [keep] constantly in movement, so as to magnify their apparent numbers."[67] German forces in World War II, for instance, used dummy paratroopers to make their airdrops seem more overwhelming, and used dummy tank bodies mounted on Volkswagons during battles in the African desert.[68] General George Kenney used several key deceptions against the Japanese in the same war, from drawing their attention to fake airfields to staging "simulated paratroop operations" in the Philippines.[69] Many other stories of deception abound, and are among the most interesting in the history of warfare.

A tactic that fits under the heading of deception is *diversion*. The feint is a tried and true tactic intended to draw the defenders' attention away from the primary objective, and therefore intended to deceive. Indeed, British historian and theorist B.H. Liddell-Hart's *Strategy*, which lays out case after case of what he called the "indirect approach," is an excellent text on how commanders have used diversions and feints throughout history.

As terrorists, guerrillas, and bellicosi progress from near-revolutionary to conventional thinking about warfare—not that they will ever fully adopt conventional methods—they will undoubtedly learn this aspect of deception. They may not mount small attacks in themselves because of their limited resources; their diversions may consist of targeted misinformation (perhaps in the form of rumors spread on the Internet or among the story-hungry media) or more sophisticated feints such as utility interruptions or computer network intrusions. Al Qaeda learned quite well the value of simultaneous attack, and used that knowledge to devastating effect in New York and Washington; they continued their penchant for coordinated attacks in bombings in Madrid and London, and in other subsequent actions. When the next bellicosi strike occurs, we should keep vigilant in case it is only a jab while a roundhouse punch is waiting to land.

Regardless of the means used, the principle of security focuses on denying the enemy an "unexpected advantage." This indicates that security alone does not win battles. With this in mind, does security really rank with the other principles, those that are directed primarily at the where (objective) and how (mass, maneuver) of warfare? Probably not. Liddell-Hart's "indirect approach" is a warfighting strategy that relies on security applications: not telegraphing your real move, making the adversary respond to different threats while you select between possible objectives. Security is an important element, but not able to stand alone: the interdependence of security with other principles makes it impossible to apply any stringent analysis. Its fate must await the analysis of interactions in Chapter 22.

Surprise. Strike the enemy "at a time or place or in a manner for which the enemy is unprepared."[70] The US Army's *Field Manual* 3-0 describes its advantages as follows: "Surprise delays enemy reactions, overloads and confuses enemy command and control systems, induces psychological shock in the enemy, and reduces the coherence of the enemy combined arms team."[71]

This principle carries at least the germ of Jomini's fundamental principle of concentration, and is another "truth" for which we have any number of examples, such as the Israelites' defeat of the Midianites—in which Gideon's picked men advanced stealthily carrying torches hidden in clay pots and then surprised their foe by smashing the pots and blowing trumpets[72]—the attack on Pearl Harbor, the German outbreak in the Battle of the Bulge, the ground sweep in Operation Desert Storm, and the bellicosi's first strike in the Terror War.

Like security, surprise alone is not meant to win battles. It is much more difficult to prevail when attacking an obvious place at a known time, but our detailed examination of this principle must wait for the discussion of interactions in Chapter 22. Indeed, like mass and economy of force, security and surprise are implicitly linked because we cannot achieve surprise without security. And, like some other principles, this one carries an unwritten corollary: preparedness, or the honest assessment of our own vulnerabilities to attack (by anything from missiles to toxins to improvised explosive devices to computer viruses).[73]

Surprise may begin a campaign, but rarely ends one; surprise can allow a single attack to succeed, but rarely carries through to a string of attacks. Thus it seems that terrorists and the genuine bellicosi rely on surprise attack after surprise attack without any obvious underlying or overarching plan; it may be too generous to ascribe to them a firm grasp on Liddell-Hart's indirect approach. For example, the Madrid train bombing was clearly a surprise attack, which marked it as more likely the work of Al Qaeda than Spain's Basque separatists—because the Basque terrorists routinely provide advance warning of their attacks. And because such attacks succeed most when they surprise, pre-attack indicators from Al Qaeda are usually few and vague.

Simplicity. "Increase the probability that plans and operations will be executed as intended by preparing clear, uncomplicated plans and concise orders."[74] It is interesting that *Joint Publication* 3-0 mentions the probability that execution will match intent, but not the probability of success.

This principle, while helpful on the surface, ultimately frustrates in practice. Wellington's written orders, penned quickly and often under fire, were eminently simple, while the battles in which he engaged were huge and complex.[75] Jomini stressed the need for simplicity ("The more simple a decisive maneuver is, the more sure of success will it be. . . ."[76]), but a truly simple maneuver—much less a truly simple battle—is hard to find.

For example, was the Normandy invasion simple? Was the "left hook" of the First Gulf War simple, or the storming of Baghdad in the second? To address even smaller actions, the April 1986 US raid on Libya, involving as it did complex air routing, refueling, and the like, was not logistically simple. In contrast to those complex operations that succeeded, the attempted rescue of US hostages in Iran was, as an operation, fairly simple to describe but difficult to execute because it was needlessly complicated by inter-service rivalries—each service insisted on having a "play," though they had little common equipment, tactics, or procedures—and the operation ended up a miserable failure. The success of Seal Team Six in 2011 showed that the lessons of Desert One had been learned very well indeed, but that action also was undoubtedly more complex than it seemed to outside observers.

The underlying concept for an attack might be simple: secure the beachhead and advance against the enemy; rescue hostages held in a foreign city; hijack commercial airliners and crash them into buildings. Every simple idea proves otherwise in execution, though, because this "principle" does not take into account the advance planning, logistical support, or other factors necessary to stage the attack; nor does it take into account the fog and friction that will develop during an attack. Thus simplicity, like so many of the others, cannot be said to explain or predict victory on the battlefield.

As we see, the nine classical "Principles of War" are important considerations and even necessary parts to conducting military operations, but they do not form a coherent theory capable of explaining past success or predicting future success with confidence. And they are not the only "principles" for commanders to consider.

20

BEYOND THE CLASSICAL PRINCIPLES OF WAR

As noted previously, *Joint Publication* 3-0 includes three additional principles in its twelve "Principles of Joint Operations": restraint, perseverance, and legitimacy. In fact, the US Air Force includes yet another additional principle in the form of "unity of effort." We will not examine these so-called principles in the same detail as we examined the traditional principles in the last chapter; they do not warrant being treated as principles, but they do raise some interesting questions.

To begin, these additional concepts were originally found in guidance for "military operations other than war," a term often synonymous with low-intensity conflict but which may have been a better descriptor than "LIC" because "MOOTW" involved more than simply battle on a limited scale. MOOTW had its own doctrine document (*Joint Publication* 3-07), though the publication no longer exists in the Joint canon; JP 3-07 included *six* MOOTW principles intended to help "ensure success and minimize losses by focusing on aspects of MOOTW that deserve careful consideration."[1] At the time, MOOTW was considered different enough in kind and character from war in general that different principles appeared to be necessary; but, in the continuing development of military knowledge and practice, MOOTW has now been discarded and three of its principles elevated in status.

It seems categorically odd that "principles" developed for something specifically "other than war" have been raised to a level of prominence equivalent to the classical principles of war. Let us consider them.

Restraint. The old JP 3-07 stated that "MOOTW may require restraint in order to apply appropriate military capabilities prudently," which was problematic because of the conditional "may." If restraint was optional, then it would not be a principle; but unless one proposed to wage "total" MOOTW, restraint would not be optional.

Current doctrine states the purpose of restraint as, "prevent the unnecessary use of force,"[2] whereas a previous version[3] prefaced that with the need to "limit collateral damage." We may accept that restraint is desirable either in terms of preserving resources or in reducing the effects of war, but whether that justifies including it as a principle of joint operations is another matter. In addition, we may question whether this principle was followed in specific instances such as the raid on Osama bin Laden's hideout, especially given the caution that, "Excessive force antagonizes those parties involved, thereby damaging the legitimacy of the organization that uses it while potentially enhancing the legitimacy of the opposing party."[4] Regardless of how we may scrutinize it, in the end "restraint" seems to be little more than a euphemism for economy of force.

Perseverance. JP 3-0 states that, "The purpose of perseverance is to ensure the commitment necessary to achieve national objectives."[5] It is unclear both how perseverance ensures commitment rather than commitment producing perseverance, and how this can be considered to rise to the level of a principle of war. It is laudable in terms of its attempt to forge a commitment to stay the course, but one wonders if it is really necessary—or if it is truly in the purview of the military in a democratic republic to take responsibility for forging such a commitment. The Joint Publication continues:

> Perseverance involves preparation for measured, protracted military operations in pursuit of national objectives. Some joint operations may require years to reach the termination criteria. The underlying causes

> of the crisis may be elusive, making it difficult to achieve decisive resolution. The patient, resolute, and persistent pursuit of national goals and objectives often is essential to success. This will frequently involve diplomatic, economic, and informational measures to supplement military efforts.

Against this guidance we may place Sun Tzu's advice that "there has never been a protracted war from which a country has benefited."[6] We may state with some confidence that a well-trained and adequately-equipped military will prepare and persevere in a just conflict as least as long as the citizenry, but especially under the U.S. tradition of civilian control it is not up to the military how long that perseverance will last.

Legitimacy. This last of the "Principles of Joint Operations" is interesting, but also appears to be less helpful than the traditional principles in planning and executing warfare. "The purpose of legitimacy is to maintain legal and moral authority in the conduct of operations."[7]

In this formulation, legitimacy seeks to capitalize on the moral force of war that USAF Colonel John Boyd considered the most important. *Joint Publication* 3-0 even alludes to this when it states that legitimacy "is based on the actual and perceived legality, morality, and rightness of the actions from the various perspectives of interested audiences."[8] Depending on the operation in question, the interested audiences may be quite broad: our own citizens and leaders, allied and other foreign governments, the populations in and near the operational area, even multinational organizations. Claiming that a course of action is legitimate can be a delicate matter, and maintaining it over the long term even harder, as

> Legitimacy may depend on adherence to objectives agreed to by the international community, ensuring the action is appropriate to the situation and to perceptions of fairness in dealing with various factions. Restricting the use of force, restructuring the type of forces employed, protecting civilians, and ensuring the disciplined conduct of the forces involved may reinforce legitimacy.[9]

It is interesting to note, as we have elsewhere, that this "principle" has evolved over time. Earlier Joint doctrine defined the purpose of legitimacy in more direct (and less conciliatory) terms: to "develop and maintain the will necessary to attain the national strategic end state."[10] The will to war is an important consideration, but legitimacy as a principle did not seem to be directed at developing that will among the US military or citizenry. It seemed to be aimed more at convincing others that military action was warranted than in determining, for instance, if forces were about to be committed to a just war.

In keeping with the importance of perception as it relates to establishing or maintaining legitimacy, this "principle" seems to rest on the earlier principle of restraint ("restricting the use of force"). Thus, not only is its status as a principle somewhat precarious, but the perception of legitimacy can never be guaranteed. Factions will almost always arise to challenge the legitimacy of a military action.

The US Air Force carries over a fourth "principle" from the Military Operations Other Than War structure: **Unity of Effort**, which comes into play "in contingency operations" when an operation is not strictly military in nature and the military is only "one agency of many" that finds itself "supporting the other instruments of national power."[11] This differs from unity of command in that disparate agencies are unlikely to acquiesce to military leadership, just as they would be unlikely to pursue strictly warfighting aims. Unit of effort, therefore, is more focused on building consensus and understanding so the participants can work toward a common aim. As with the rest of the principles of Irregular Warfare, however, it is not useful as a theory.

Other candidate principles. In a 2012 article, Army Lieutenant Colonel Tom Kratman proposed three additional principles which have much more potential than do restraint, legitimacy and even perseverance: Attrition, Annihilation, and Shape.[12]

Attrition recognizes that, when it comes to waging war, "everything costs and nothing is cheap." To be prepared to fight means having resources that will be expended in the effort, because it is not a matter of whether losses will occur, but when. By extension, being prepared for those losses means considering whether and how they can be replaced and how costly that will be in time or treasure. Kratman likens this principle to "the British principle of 'Sustainability' and the Soviet principle of 'Logistics,'" but notes that those others "presuppose adequacy of available replacements and replenishments." In the confusion and chaos of the crisis, however, replacements may not be forthcoming; depending on the nature of the engagement and the difficulty of resupply, replacing losses may range from merely inconvenient to absolutely impossible. Over time, unit effectiveness, efficiency, and morale will diminish along with supplies.

Annihilation, which Kratman introduces as a Russian principle of war but which has also been central to what Russell Weigley called the "American way of war," is related to attrition in that it seeks to degrade the enemy's ability to continue the fight. Its effect is certainly similar to attrition, though it achieves that effect differently and ideally on a much shorter time scale. As Kratman notes, "when an enemy force is completely and utterly destroyed, annihilated, it cannot be rebuilt in as effective a form, if indeed it can be rebuilt at all."

Shape, which Kratman also refers to as Geometry, "concerns the physical shape of the ground, . . . in relation to enemy forces and weapons effects, time-space relationships, modified by transportation nets and obstacles, and the like." Unlike the generic view presented by Sun Tzu's "Nine Grounds" in *The Art of War*, this principle takes into consideration the terrain and other features of a specific battleground (or battlespace) being contested. The lay of the land, avenues of approach, obstacles, sight lines, fields of fire, and the like are all part of the "chaos of the given" which the commander must come to understand—and to which he must adapt.

Kratman illustrates the concept using the 371 B.C. battle of Leuctra, in which "the Thebans under their commander, Epaminondas, [set up with] a very strong force on the left, a weaker force in the center and set back, and a very weak and thin force on the right, set back still further." Opposite the Thebans, the Spartans "set up in the usual broad phalanx." As Kratman wrote,

> What happened on the field was that the Thebans' strong left struck the Spartan right with overwhelming power, crushing it before their own much weaker center and right could be engaged. The military

term of art for this is "echeloning." Now imagine it without "Geometry" or "Shape" with the whole Theban army organized the same way but all on line. Correct: The strong Spartan left and center crush the Thebans opposite them, setting the Theban army to rout, before the Theban left can be decisive.

There is no denying the influence that Shape can have on a battle, but since it depends on the physical reality presented to each side it seems to form part of the "chaos of the given" that each commander must consider and interpret using the theories they bring to the fight. As such, while it is a useful concept and may even warrant consideration as a principle, it probably cannot qualify as a theory such as we have been discussing.

Indeed, none of these additional principles appear to form any more cogent theory than did the traditional principles, and only a few of them add much to the usefulness of the nine classic principles as a whole. If nothing else, these additional principles illustrate that ideas come into and fall out of favor from time to time, as seen in the name change from MOOTW to Irregular Warfare and the way the concept of legitimacy shifted from developing the will to achieve objectives to building a moral foundation for action. And as Kratman noted, all principles—including his own—are not "checklists to victory" or "formulae for success," but instead serve as texts for studying war, frameworks for structuring training, and "warnings against nonsensical theories."

Perhaps it may be said that they warn against everything in this book.

21

THE POSSIBILITY OF ERROR

From the last two chapters we can state with some confidence that the traditional or lately-added "Principles of War," taken individually, lack much in the way of power to explain or predict the success or failure of military operations, which we have stated is something we expect of a sound theory. Examining each principle revealed either specific contrary examples or sufficient probability of exception to call it into question. Perhaps instead of principles, we should refer to them merely as "considerations" of war.[1]

Or perhaps we should not refer to them at all. Another point to remember about the Theory of Knowledge is that knowledge is probable only; i.e., all knowledge is susceptible to error, including the foregoing analysis.

You may have noticed that our approach to each principle was to build mentally and verbally what may be visualized in a modified Johari window.[2] Illustrated in Figure 5, on the left side we placed the theory or principle in question, and whether it was applied or not; along the bottom we placed two simple outcomes—victory or defeat. For a principle to qualify as a good theory under this analysis, its presence had to explain or predict victory, its absence loss; the "grey areas" should be devoid of examples.

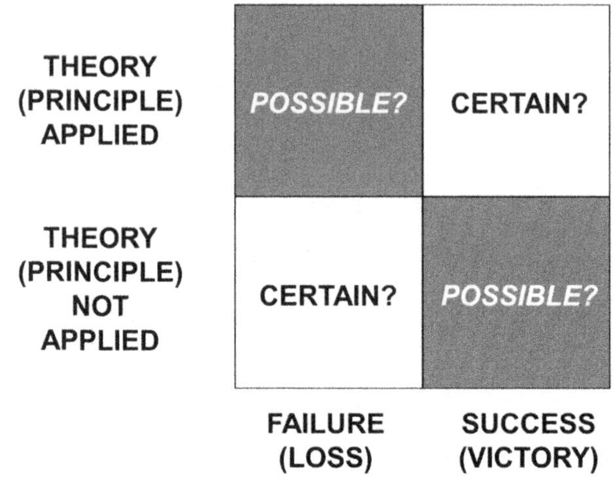

Figure 5. Theory described in a modified Johari window. Purists such as Jomini assert that applying a principle or principles makes success certain; our investigation focuses on the "grey areas" of uncertainty.

But as we saw when we looked at the principles, reality is not so clean and simple. Successes gained without invoking a principle, or losses when the principle was in effect, fill those grey areas. Theory of Knowledge requires that the theory—in this case, the principle—be either rejected or modified to take into account those instances. The historical approach, in contrast, has been to let the principles remain more-or-less intact, and either emphasize or de-emphasize them instead of revising them. The end result is that the principles continue to advertise common-sensical items, such as unity of command, the truth of which seem self-evident but the application of which leave grey areas of doubt.

But this success-or-failure model is too restrictive. A better approach would be the one presented in Figure 6, where the application of theory and the resultant outcomes are shown as two continua. Applying this model, a principle's strength as an evaluative or predictive theory would correlate to its degree of success. This analysis is more complex, since differing degrees of each item must be considered. We will discuss that later.

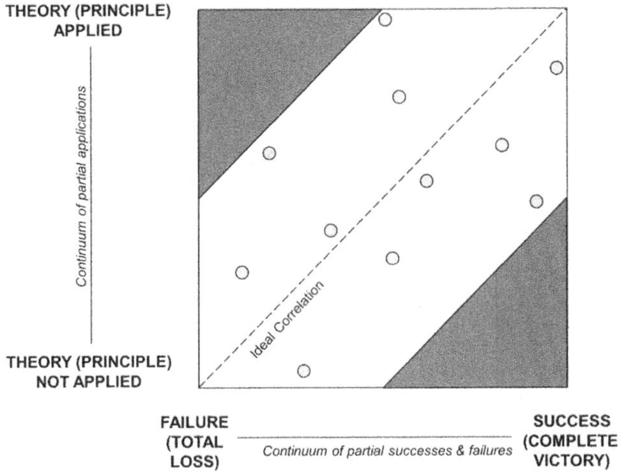

Figure 6. Correlation of theory and success along two continua. The closer each individual example is to the ideal case, the better the theory, although correlation does not automatically indicate causation. (The apparent size of the "grey areas" is irrelevant since this illustration is entirely qualitative.)

Our analysis of the principles is also overly simplistic for another reason. We examined each principle on its own merits, but treating them individually is inherently weak. As discussed several times up to this point, the principles do not appear or act alone; this is why Air Force doctrine states, "No one principle should be considered without due consideration of the others."[3] To give due consideration, we will begin examining interactions between them in the next chapter.

With respect to the conclusions offered so far, Jominians may rally around his statement that "a single exception cannot disprove a rule based upon the experience of ages and upon natural principles."[4] Perhaps not, if the "rule" is not really a rule *per se*; however, any exception can cast doubt. But as discussed many times, the epistemology presented earlier requires that the exception be explained and the explanation made part of the theory. Jomini made no such attempt; rather, he used carefully chosen campaigns to illustrate his central principle, admitting there were exceptions but not considering them important: "We will . . . show, by the history of twenty celebrated campaigns, that, with few exceptions, the most brilliant successes and the greatest reverses resulted from an adherence to this principle in the one case, and from a neglect of it in the other."[5] Indeed, Jomini called the exceptions "cases from which nothing can be inferred in the face of thousands of opposed instances."[6] While our analysis is not yet complete, hopefully we can avoid falling into that error.

The relationship of military history to theory of war was described well by Clausewitz, who argued against the reduction of theory to principles: ". . . our historical survey can end. Our purpose was not to assign, in passing, a handful of principles of warfare to each period. We wanted to show how every age had its own kind of war, its own limiting conditions, and its own peculiar preconceptions."[7] Perhaps this is why late in the 20th Century the principles fell somewhat out of favor; for example, only being taught at West Point as part of one "thread of continuity."[8] Perhaps this is also another reason for recent expansion of the principles with "restraint," etc.

We must point out that this is not the first argument against the universality of the principles. They have been challenged many times. As noted, they have also been changed before, which accords well with Theory of Knowledge but violates the idea that they were supposed to be unchanging. At other times they have been discarded entirely.[9] We may therefore expect to see the principles fall into disfavor and recover again and again as new technologies and methodologies come into play, even though arguments against them only earn us Jomini's pity: "If a few prejudiced military men . . . still contend that [war] has neither principles nor rules, I can only pity them, and reply, in the famous words of Frederick, that 'a mule which had made twenty campaigns under Prince Eugene would not be a better tactician than at the beginning.'"[10]

The principles may not satisfy as a theory, but do we have enough reason to completely discount them? Not at all. They are still useful—and in combination all the more powerful than shown above. The partial theory they represent may be modified or reformulated to better explain the past and predict possible future outcomes.

22

COMPLEXITIES AND INTERACTIONS

Examining the Principles of War as individual theories without taking into account the effects they have on each other was, as we continue to stress, overly simplistic. At no point did we argue otherwise: we mentioned several times so far the weakness of examining individual issues one at a time. We took that more direct tack, however, because combining multiple theories requires careful consideration of how the theories interrelate. Indeed, the difficulty of treating all the principles together will become clear very shortly. But it is vital to study their interactions; as Clausewitz noted, "effects in war seldom result from a single cause; there are usually several concurrent causes."[1]

We should also note that Clausewitz maintained that "each identifiable cause . . . has to be correctly assessed," a decidedly Newtonian outlook.[2] That Clausewitz would view the world in Newtonian terms—clear cause-and-effect, every action with its equal and opposite reaction—is not surprising. Nor is it surprising that we, even with the advanced scientific knowledge we have learned, view the world in the same way: most of the phenomena we encounter in our daily lives appear to fit into the single-cause-and-single-effect mode. But appearances can be deceiving.

Air Force doctrine notes that the principles of war "are independent, but tightly fused in application."[3] The older *Basic Aerospace Doctrine* put it this way: "All of the principles are interrelated and interacting elements of warfare. They are not separate and distinct entities from which a commander selects when employing forces."[4] Recall that the old *Joint Publication 1* noted that the principles "are in some cases paired juxtapositions"; it warned commanders to "exercise judgment in applying these principles."[5] In truth, the situation is worse than simply pairs of principles: examining all the possible interactions between the principles, in order to determine the optimum "mix," is extremely complex.

When many different inputs combine in a complex system to produce a single outcome, it can be difficult to identify which input was the driving factor. Interactions between the inputs often influence the overall result more than do the individual inputs themselves.[6] That is, individual factors (e.g., security, surprise, etc.) influence the outcome, but those individual factors also influence each other. We may think of any number of possible examples. An offensive's success may be influenced by whether it is kept secret. The ability to maneuver effectively may depend on the clarity of orders and the straightforward nature of the command chain. This is the nature of interactions, which influence events around us but which we may not always recognize in our preference for simple Newtonian causes. As noted by eminent statistician Don Wheeler, "Two factors are said to interact when the effect of one factor upon the response changes depending on which level of the other factor is present. . . . When two factors interact, it is impossible to talk about their effects upon the response separately."[7] More precisely, it is impossible to explain their effects without considering how the factors interact.

To demonstrate this phenomenon and determine what interactions drive system performance, engineers and systems analysts often perform designed experiments on complex systems. In classical design of experiments, investigators find out which interactions are dominant by systematically accounting for all the possible combinations of variables.[8] In like manner, we might try to demonstrate the interactions between all the principles of war, taking each combination of the nine principles (or twelve, or whatever number one chooses) as a single case until we have studied them all. However, a little mathematics shows that the effort is far beyond the scope of this or probably any work.

Consider briefly the case of a pair of factors, each of which may either be present or absent. This is illustrated in classic form as a two-by-two (2x2) experiment, and has four distinct cases; an example is found in Table Two, in which the factors are types of rabbits, either male or female. The presence or absence of each factor is indicated by the number of each type, zero or one, placed in a cage. The outcome, shown in the shaded area of the table, is the total number of rabbits in the cage. The first, or "null," case has zero male and zero female rabbits in the cage, with the obvious result that the total number of rabbits is zero. The "experiment" continues through all the possible combinations of the factors, and illustrates the powerful interaction between the factors, i.e., between the one each male and female rabbit. Mathematically, the number of interactions is determined by taking the number of variations being considered (in the example given, presence or absence representing two levels at which the factors may be set) to the power of the number of factors (male or female rabbits being two factors): 2^2 equals 4.

Table 2. The Power of Interactions. The 2x2, or "Rabbit," experiment (as described by Dr. John Surak, Clemson University).

Rabbits	0 Females	1 Female
0 Males	0	1
1 Male	1	Many!

The interactions between the nine classical principles of war are not so simple. If each is considered in binary terms—applied or not; or, measured as high or low values—the resulting number of interactions is given by two to the ninth power (2^9), for 512 interactions. While this seems high, it is actually too restrictive, since a better model would have the principles operating along continua (as presented in Figure 6). A continuum has an infinite number of possible values, but if we consider only three (being high, moderate, and low levels), the number of interactions grows to 3^9, or 19,683 interactions!

This assumes, of course, only the nine classical principles—but as discussed previously the number of principles is not static.[9] If only the suggested principles of flexibility and logistics (mentioned in chapter 18) were included, the number of binary interactions would be 2^{11}, or 2048, and the high, moderate, and low analysis would produce 3^{11}, or 177,147 interactions. If the twelve current principles of joint operations were included, the number of interactions would be 2^{12} (4096) or 3^{12} (531,441), respectively. It is also possible that the number of principles might be different for different services: as discussed, the Air Force includes unity of effort among its principles, and for a time included three additional "airpower" principles in its doctrine—timing and tempo, logistics, and cohesion—though they were later discarded because they lacked the "authority lent by familiarity."[10]

Beyond the fact that more principles mean many more interactions between them, the case of the discarded airpower principles raises two issues. If the explanation is taken at face value, the additional principles were rejected because the others were more familiar and therefore considered more authoritative; however, we may ask if familiarity is a valid reason for accepting a theory? For many years, people were familiar with and accepted the theory that the sun revolved around the earth, but that did not make the theory correct. (In fact, it required the introduction of additional subordinate theories such as "epicycles" and "equants" in order to explain observations of the moving planets.) Second, though Joint and Combined operations require intimate coordination and knowledge of service procedures and protocols, and our experience in Gulf War Two and afterward showed that the US military has made tremendous strides in this area, we still struggle to some extent with "jointness." Why then should we expect the largely Napoleonic and ground-based traditional principles to be valid in the modern age of combined operations? Today, for instance, the US Air Force includes seven "tenets of airpower" as adjuncts to the principles: centralized control and decentralized execution; flexibility and versatility; synergistic effects; persistence; concentration (which recalls, even in its discussion, mass and economy of force); priority; and balance.[11] This raises the question whether, in the evolving analysis, any of these other factors might ever rise to the status of "principles." It is not as far-fetched as it might seem, given the changing nature of warfare.

To return to the topic of complexity, it should be clear that a comprehensive study of even 512 interactions between the classical principles is far beyond the scope of this monograph; since many of those interactions are likely to have significant effect, however, it should be equally clear how important it is that we not neglect them. As Clausewitz wrote,

> . . . success is not due simply to general causes. Particular factors can often be decisive—details only known to those who were on the spot. There can also be moral factors which never come to light; while issues can be decided by chances and incidents so minute as to figure in histories simply as anecdotes.[12]

Or, they can remain unknown forever.

23

FROM ANALYSIS TO SYNTHESIS

Democritus, the ancient Greek philosopher and scientist, postulated that by subdividing a piece of material over and over one would arrive at a final bit of the matter that, were *it* to be divided, would lose the characteristics of the material. He called this final bit the "atom." As time progressed, molecular science gave way to atomic, and atomic science to sub-atomic, as we discovered more about the structure of the universe. If we examine the similarities between the "Principles of War," will they lead us to smaller, more fundamental building blocks of successful conflict, or to more complexity than we imagined?

Based on the mathematics in the last chapter, we established that the interactions between the nine traditional principles form a construct too complex to consider in total here. Therefore, even though simplification has its own hazards, it might be to our advantage to reduce the number of salient principles to something more manageable. In order to do so, we will need to move from analysis to synthesis: from breaking war into its constituent parts and examining those parts separately, as we did with the principles, to putting the parts together and examining the results. This should give us fewer pieces to consider, and therefore enable us to study the interactions between them more easily. Unfortunately, like five-year-olds examining an old spring-driven wristwatch, we may find that taking things apart is easier than putting them back together.

Traditional warfare, i.e., the major clashes between nation-states that characterized warfare for thousands of years, seems to take the form of mass + maneuver + objective. (The notation is a convenience only; addition is not really the operation in action here.[1]) We already stated at the outset that an objective is necessary to any conflict, and this formulation falls directly in line with Jomini's prime principle. This combination is not at all new, of course, but it illustrates the idea of putting the principles together to see what they produce. And it can be extended, by including two additional principles to produce Liddell-Hart's indirect approach: objective + surprise + security + mass + maneuver.

Consider another possible formulation: security + surprise + economy of force. This formula seems to describe revolutionary warfare, i.e., the struggle of the underdog faction that does not possess the military power to confront those in power directly. For instance, some kinds of guerilla warfare can be characterized by risking as little force as possible in surprise engagements that may reap significant results. This combination may also apply to terrorism and the actions of the international bellicosi.

Now that we have seen some possible combinations and relationships between the principles, we may examine other potential new formulations in a mental exercise not unlike fractional distillation in chemistry: We apply the heat of theory to the raw material of fact, with the aim of driving off impurities while we collect the useful distillates. If what we discover is not immediately useful, perhaps the effort itself may have some instructive value, since even if it does not eventually lead to *coup d'oeil*, it forces us to think about familiar subjects in new and different ways.

Synthesizing new constituents or even new principles may involve more ideas than those we have already considered. In much the same way that past theorists have ventured outside the sphere of war-making for new or unique ideas that may be applied (recall that Clausewitz used the natural phenomena of fog and friction, and Warden applied systems theory to describe the enemy), we may be able to find useful components outside the military world. Whether they fit well or poorly remains to be seen; hopefully if they are not perfect fits it will not be too hard to shave off the square edges of the pegs in order to work them into the round holes.

Consider, for instance, another facet of systems theory than Colonel Warden did. We may think of war not as a complete and self-contained whole but as a series of interconnected and interdependent actions. It certainly is exactly that, and the relationship between tactics and strategy may be thought of as the process of ensuring that all the disparate actions further the overall war aim. In other words, war itself may be considered as a *system*—an interconnected set of processes—and systems analysis techniques may be helpful to understanding it.

If war *is* a system, one approach to analyzing complex systems that may help analyze war is the "fault tree." The fault tree methodology is often used in systems engineering to evaluate the reliability of a complex system, or to investigate the failure of one. The method involves a detailed graphical representation of system processes and components, noting cause-and-effect as a function of whether separate events must happen together to cause a failure or if items may cause failures by themselves. Analysts often combine this graphical representation with estimates of the probabilities of the postulated failures. For especially complex systems, the tree will be proportionately complex: for example, "more than 3000 individual elements in the [Space Shuttle] *Columbia* accident fault tree were examined" in the investigation process after its reentry failure.[2] In many such cases, where actual graphical trees are too complex to be useful or simply too time-consuming to produce, the events and nodes may be translated into Boolean algebra and evaluated symbolically. The end result should be the same: identification of failure points in the system and the pertinent inputs to potential (or actual) catastrophic failures.

In the next chapter, then, we will consider how war can be described as a system. Subsequently, we will apply the fault tree methodology to see what that produces.

24

THE SYSTEM OF WAR

It may seem odd to categorize war, which is not a discrete thing but rather an abstract notion describing events, as a system. A system, and especially a human-designed and -constructed system, is usually understood to be a collection of things—material objects or their software-coded equivalents—working in concert to achieve some aim.

One definition of a system, which comes from a definitive text on the fault tree approach to systems analysis, is "a deterministic entity comprising an interacting collection of discrete elements."[1] This is a rather high-sounding definition; a more common understanding of a system is a collection of interrelated and interacting parts that operate together toward a common purpose.

A box of odds and ends is not a system; nor is a box of computer components until those components are assembled in working fashion. It seems that such a definition would not describe an abstract notion such as war. It could describe the military, perhaps, or organizations and mechanisms built for war, but not war itself. Or could it?

Before we attempt such a description, consider some of the difficulties of the pursuit. Critics of the attempt to treat war as a system might agree with the author of the *Fault Tree Handbook* that "It is completely futile to attempt an analysis of something that cannot be clearly identified."[2] Yet we attempt analyses of things every day which we have not clearly identified: that is part of sifting through the "chaos of the given" which we discussed with respect to Theory of Knowledge itself. If we take the quote to mean a *complete* analysis, we might agree with it; and war certainly is too complex to be completely analyzed and understood. But a complete analysis is not what we are attempting.

Our purpose is not to apply any single methodology to break down war into its component parts, but to understand more of the whole by using a variety of different methods. By way of analogy, we can compare the art of war to the art of painting. In the case of historical wars, the painting is complete (though we may occasionally encounter a forgery, a reproduction, or a hidden masterpiece); in the case of current wars, it is being painted even now. We evaluate the paintings to determine if they are masterpieces—or if they even qualify as "art." We must investigate light, shadow, color, and texture to practice our own art, but we need not chemically analyze the paint to learn what makes it burnt umber; instead, we consider the painting as a whole.

Of course the *equipment* of war may be thought of as systems: as assemblages designed and built with clear purposes in mind, they clearly fit the commonsense definition. An M16 or a Merkava or a Messerschmidt is certainly a system; an aircraft carrier with all its aircraft is a system; a carrier battle group or a constellation of satellites is a system. All of these may be grouped together in larger and even more complex systems. But what defines each as a system is the purpose for which they were assembled: the application of the individual weapons for the purpose of attaining our objectives. (This is not the old argument of "a weapon unused is a useless weapon," for the simple reason that the *threat* of violence may be sufficient to achieve the objective. Some weapons, e.g., the vast majority of nuclear weapons, fulfill at least part of their purpose by being poised for attack rather than being used in attacks.)

Still another criticism of this approach may be that the boundary of the "system of war" is not very well defined—or easy to grasp.[3] The boundary of a tank is very evident, and the boundary of a tank battalion can also be identified. But where is the boundary of war? Certainly it extends beyond any immediate battlefield to encompass the political aims of the societies waging it. Therefore we must ensure that our systems approach takes into account the larger issues and impacts of the war.

To understand any system, including understanding war as a system, we must understand its purpose, its parts, its inputs, and the interrelationships between the parts. Furthermore, especially in complex systems, it helps to understand the relationships between the individual parts and the system's inputs and purpose; this falls into the category of control and feedback, which in military systems is best labeled as "command." It is important to note that the system must be integral; in other words, "the properties of any system are lost if it is disassembled."[4] Take a helicopter, for example: It is a collection of specially assembled parts that work together to achieve the purpose of powered flight. Remove the engine from the helicopter and it is no longer, from a systems standpoint, a helicopter. It may *look* like a helicopter, but it cannot *function* like a helicopter: as a system it is incomplete. And if the helicopter engine itself is adapted in some way to provide power to another type of vehicle, then at that point it is no longer, from a systems standpoint, a helicopter engine.

For the system of war, the purpose is to achieve victory (i.e., to seize the objective) by force or by the threat of force. It is important to note that force is required not because it is desired or desirable, but because it is needed to counter the opposing force. The pieces of the system may be stated simply as the enemy and ourselves (with allies, as applicable), or may be taken as a complex whole of the various units involved in the action. The inputs to the system are the personnel, equipment, and methods both sides bring to bear—recognizing that the entire system is built, maintained, and used within the context of society, i.e., the attitudes and will of the people and the make-up of the military.

Discrete, tangible items are certainly easier to conceptualize as systems than abstract, human constructs. A radio, for example, is clearly a collection of parts that work together to convert electromagnetic radiation into sound waves (or vice versa). Not only can we conceive of higher level "systems of systems" all working in concert, but we can work backward from the system itself to the original, basic inputs that make it up. If the system being examined is a tank, it would have several subsystems (e.g., power train, main gun), which are each made up of specialized components (e.g., engine, transmission; fire control computer, barrel); in turn, the components are built from parts (e.g., turbine, gears; integrated circuits, input/output devices), which are themselves derived from much more basic items (e.g., steel, silicon, electricity, know-how). A system of war, if one can be imagined, must follow a similar pattern.

Following the ease with which physical items may be examined according to systems theory, it is even easy to apply systems theory to a "military system" in terms of organization, people, etc.—although equating individuals with parts/components can be dehumanizing. (Such is the nature of all things military: with respect to the friendly side, to strip away enough of the sense of individuality that the member integrates more completely into the unit; with respect to the enemy, to de-emphasize their humanity to make it easier to dispatch them.[5]) We do refer from time to time to the military "machine," however, which can lead to some interesting if debatable analogies and which might in itself be interesting to investigate in these systems-analysis terms.[6] Table Three lists some possible analogies for a "system of war" and shows that, for this approach, we may find that the operational level is useful, even if it is not, as discussed previously, precisely "real."[7]

Table 3. System of War vs. Mechanical Systems. Possible analogies.

THIS LEVEL:	MAY EQUATE TO:	OR PERHAPS TO:
Grand Strategic	System	System's Effects on Other Systems
Strategic	Subsystem	System
Operational (or Tactical?)	Component	Subsystem
Tactical (or Individual?)	Part	Component

It is not as easy to apply systems theory to an event such as a battle, except with respect to the organizational and physical systems participating in the battle. It is harder still to apply systems theory to war as a totality.

One reason it is harder to apply systems theory to human systems is that their aim may be common only temporarily; i.e., because humans (either individually or as groups such as nations) may decide that their best interests have changed, they may stop working toward what was once a common purpose. Politically, this may take the form of refusal on the part of some to support the war effort—a sometimes unfortunate tendency in democracies, where fringe groups may prefer momentary political advantage to long-term national interests and security. Militarily, this can take the form of shifting alliances or the refusal of allies to support a coalition effort.

Every system, then, must have an aim that the component parts work together to achieve. We may take the aim of the "war system" from Clausewitz: to secure the objectives of the *polis* (i.e., to continue the political discourse by military means). This is where Colonel Warden's "Five Rings" theory (discussed in Chapter 17) may fall a bit short, as it does a good job of mapping out the component parts but does not delve into the enemy's purpose(s).

The component parts of a system of war are more difficult to assess, and even naming them does not come close to examining the relationships between them. If we were starting at the grand strategic level, we might first list the instruments of power—diplomatic, informational, military, economic—though we may find that we need to add subsets as well, such as political, sociological, psychological, and technological power. Alliances would determine part of the make-up of the system, but within that broad category we might also need to consider the common, complementary, or contradictory goals among the allies, as well as any animosities that may linger between them. Relationships with neutrals may need to be considered as well. Other intangibles would have to be included, such as situational awareness, intelligence, and observations; knowledge of and predictions about the enemy; and our own hopes and fears (i.e., our "will").

Perhaps it would be better to start with a more limited example, and think of war as a system less like an artifice and more like a process in which inputs are converted into outputs. As discussed above, the inputs are still men (in the generic sense), materiel, and methods; in other words, personnel organized, trained and equipped to fight, along with the doctrine and tactics, techniques, and procedures that inform their fighting. The processes involved would include observing and assessing the enemy; predicting (on either a tactical or strategic scale) likely enemy actions; deploying and supplying troops; engaging; and eventually assimilating the gains or withdrawing, hopefully victorious. ("Observing," in this case, recalls Boyd's OODA loop, which can be used to describe the command activities involved in complex actions that require coordination of many forces and efforts, but is not as immediately useful in cases where the "orient" and "act" stages require prior planning and extensive coordination.[8]) The output of the process, ideally, will be victory—however we have defined it, given the objective we sought. Setting the objective and making the requisite strategic predictions are the higher-echelon command functions that provide control and feedback to the system.

Do not lose sight of the fact that the objective of whatever martial system we construct should be victory. Failure to keep that objective in mind is usually the fault of the political rather than the military machine. Since the mid-1980s the US in particular has searched for "exit strategies"[9] too vigorously, when it should have searched for victory strategies. The shift in emphasis from achieving full victory to being satisfied with orderly withdrawal seemed to spring from the US defeat in Vietnam, and while much of the military's tactics and organizations have improved since then, this strategic tendency has not abated.

It may be more precise to state that the military's pursuit of victory has usually stopped at the tactical and operational level; if it has approached the strategic level, it has certainly not proceeded to the grand strategic level. The campaign in Iraq springs to mind as a textbook case: the coalition achieved so rapid a military victory—a rout so complete that it surprised nearly everyone—that its overall progress toward complete diplomatic, political, and economic victory stalled. In some circles, that constituted a failure that had to be rectified by quick withdrawal, no matter what chaos may ensue; that promise helped propel Barack Obama to the Presidency, but, as with many such things, President Obama found that matters were not as simple as they were when he was a candidate. We should not be content to stop at a quick military victory unless we are reasonably sure that victory will gain us the long-term, overall victory we really need; however, we cannot know what that overall victory should look like if we have not taken the time to define it and figure out how to achieve it.

It is important to remember that, "there are degrees of victory, some better than others."[10] Planners and commanders might consider using the Victory/Defeat Space model shown in Figure 7 to determine the shape of the victory to be sought. By deciding beforehand the definitions for the minimum acceptable victory, the maximum anticipated defeat, etc., decision makers would not only approach any coming war with open eyes but may also be able to discern ways to move from the potential for defeat to the probability of victory.

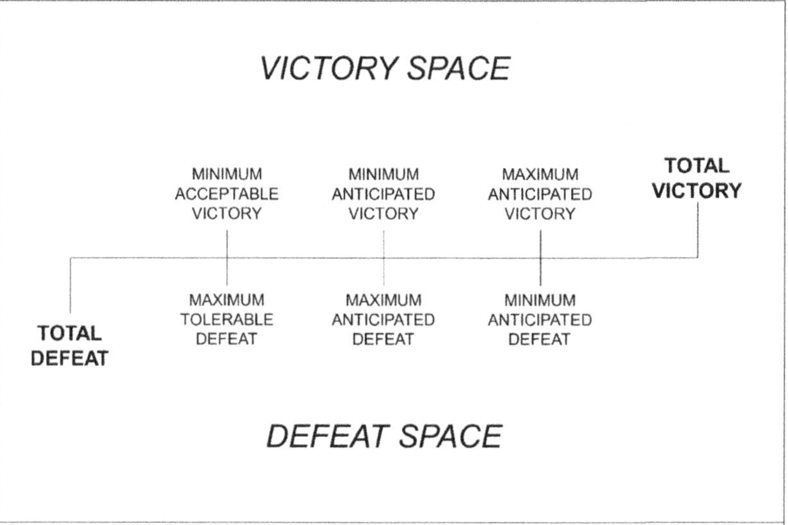

Figure 7. Victory/Defeat Space. Adapted from W.E. Vesely, et al, "Fault Tree Handbook," NUREG-0492, Figure III-1.

For example, consider a scenario that has concerned and continues to concern many analysts: China attacking Taiwan to bring it back under control of the mainland government. We may have several potential responses to such a move, and each potential response would have several potential countermoves by Beijing, but let us consider an immediate US military response that leads to a Chinese declaration of war and attacks on US bases in the region. What would the Victory/Defeat Space look like for this situation? Complete defeat for us might involve China taking US territories such as Guam or conquering US-allied nations such as South Korea. Complete victory for us might involve a Chinese retreat to its original borders along with suitable reparations. Or we might define complete victory as the dissolution of the Chinese government and replacement with a US-friendly regime; in that case, the Chinese retreat might fall into the category of "minimum acceptable victory." Our definition may, in fact, change as the conflict unfolds. And how we define the victory we want will determine the resources and tactics we need to prosecute the war—no matter what that war may be.

25

THE WAR TREE

By looking at war as a system, we may apply the Fault Tree approach to see if it gives us any insight into preparing for or prosecuting war.

Let us begin by noting that the *Fault Tree Handbook* warns, "Generally speaking, Fault Tree Analysis addresses itself to the identification and assessment of . . . catastrophic occurrences and complete failures."[1] That is, the fault tree begins with "an undesired state" or event, and analyzes a system to "find all credible ways in which the undesired event can occur."[2] The fault tree applies to discrete systems, and starts with the assumption not only that the system exists but that it can be described (if not understood) somewhat completely. Thus, to attempt this type of analysis we must accept that war can be described (if imperfectly and perhaps uncomfortably) as a system. Systems engineers often use the methodology to find single points of failure that are likely to cause the undesired event to occur, because they can improve system reliability if they can reduce or eliminate those critical failures.

What is the undesired event in this application? War itself is one option, since we much prefer a prosperous peace. But if war is the system (as discussed earlier), then the undesired event is defeat. We will examine both options in this chapter.

First, let us consider war itself as the undesired event. Figure 8 shows the beginnings of a fault tree in which peace is the "normal" or at least the desirable state, and war the fault. This "war tree" shows the first stage of an analysis of the inputs that would produce the war. In this case, the system that breaks down because of the fault is the larger geopolitical system, at which point an enemy—either a single nation, an alliance of nations, or a sub- or supra-national group—confronts us. As in Chapter 8, the other two inputs necessary to produce war would be violent action, for which both arms and the will to use them are necessary inputs, and a national objective for which we must struggle against the enemy.

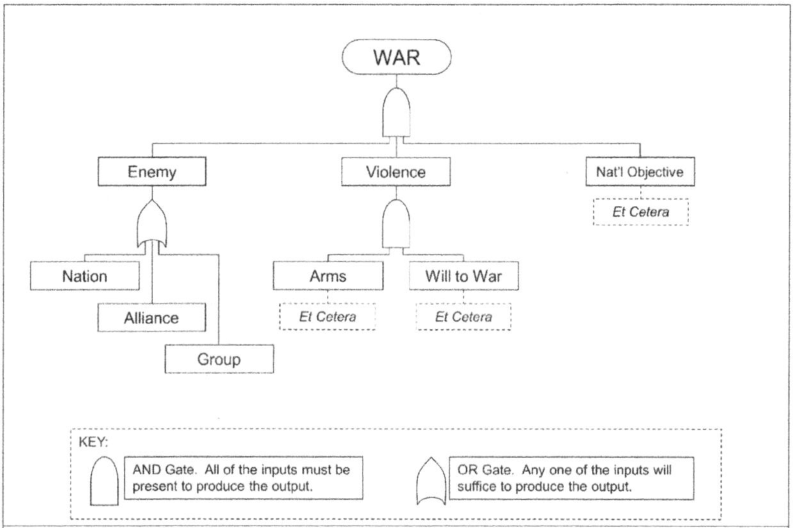

Figure 8. War Tree General Example. The AND-gate at the top signifies that all three of the inputs must be present to qualify as a war. The OR-gate under "Enemy" signifies that any of the listed inputs may produce an enemy; until very recently, "group" would not have been included.

The figure shows that some inputs can cause effects on their own, while others must work together. Specifically, the enemy may be either a single nation, an alliance, or in this age of the bellicosi, a smaller group; in contrast, the violence or threat of violence must consist of a combination of arms and the willingness to use them. How deep we delve into the analysis depends on how much time we have, and how much fidelity we need. At first glance it might seem that expanding the "arms" input would require, at a minimum, that we assess our order of battle, our industrial base, acquisition practices, and so forth. However, the analysis could take another turn: raw materials, industrial capacity, wealth, scientific creativity, etc.

Now let us consider war itself as the system being studied, and defeat as the undesired event. In this case, we must postulate the possible "inputs" to that defeat. At the same time, we must consider their relationships to one another—i.e., those inputs that act independently and those that must happen sequentially or simultaneously. Figure 9 shows the beginnings of this war tree, with the interconnections between the elements that make up the war or battle.

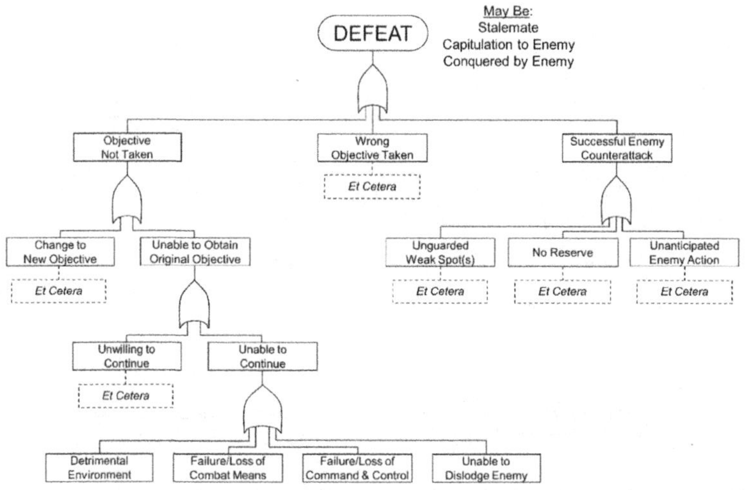

Figure 9. War Tree General Example II. Examining possible inputs to defeat as the "fault" in the system of war.

In Figure 9, the defeat possibilities—stalemate, capitulation, and conquest—correspond roughly to the maximum anticipated, maximum tolerable, and total defeat spaces, respectively, from Figure 7. (We may prefer to consider capitulation as a total defeat, especially in smaller conflicts, in which case a stalemate would be the maximum tolerable defeat; however, making that determination would be a separate matter from examining the War Tree idea itself.) In this abstract example, the defeat is characterized as being caused by either a successful enemy counterattack, the failure to capture the objective, or capturing the wrong objective; in concrete examples, the inputs to defeat may be vastly different.

This type of War Tree has an advantage over the tree shown in Figure 8, in that it allows for consideration of much more concrete elements. Digging deeper into the potential causes of defeat—i.e., expanding the tree to lower and lower levels—may reveal the combinations of enemy action and friendly mistakes that are most likely to produce defeat. Doing so in real-world cases would require breaking down historical battles to a greater degree than we are prepared to attempt in this text: while we might rely on previous analyses of historical battles to build a War Tree to illustrate, such an examination will have to wait for another opportunity. Nevertheless, it seems that understanding the combinations of inputs and their interdependencies may allow us to take steps to reduce the probability of events likely to produce defeat—or at least make us better prepared to deal with their effects.

These general examples illustrate the problem we encounter when we fail to establish the system boundaries tightly enough. Using the War Tree to examine too large a scenario would be akin to using a magnifying glass to study the Rocky Mountains. This tool is better suited to small examples with clear boundaries. A more specific example of the War Tree as it might apply to missile defense is shown in Figure 10. Here, identifying each of the possible inputs that would result in a successful enemy strike serves a critical purpose: it focuses attention on those areas that need to be most robust in order for the missile defense to work properly. Figure 10 is still quite a high-level examination, though. To have the maximum benefit, we would have to carry the analysis further into the innermost workings of the missile defense system itself, to identify the subsystems (i.e., the hardware) that might contribute to the failures; any weaknesses in the tactics, techniques, and procedures which will govern the system's operation; and any enemy capabilities that may interfere with our missile defenses.

It is possible, though more difficult, to turn the focus of the fault tree around by searching for the inputs to a desirable rather than an undesired event. In this case, of course, it is no longer a "fault" tree; it may be considered a force tree. Were we to take this approach with the War Tree, the top-level event would change from "defeat" (as in Figure 9) to "victory," and we could build a tree that examined those inputs/events we think are necessary to bring that victory to pass; e.g., establish air supremacy, capture territory or other objectives, prevent resupply of the enemy, etc. The result would look comprehensively at possible inputs to victory, but to carry on such an analysis beyond perhaps two or three levels would lead to much more complexity. It would soon reach a point of diminishing returns. (On the other hand, the end product might be thought of as an active organizational tasking chart, where subordinate commanders at each lower level were charged with providing the necessary services/inputs, all of which would roll up into preparations for victory.) We will not explore this possibility here because the approach entails three primary weaknesses: the difficulty of accounting for all of the factors that would contribute to the outcome; the impossibility of including potential (or actual) enemy actions that would affect the overall probability of victory; and the inability to account for fog or friction.

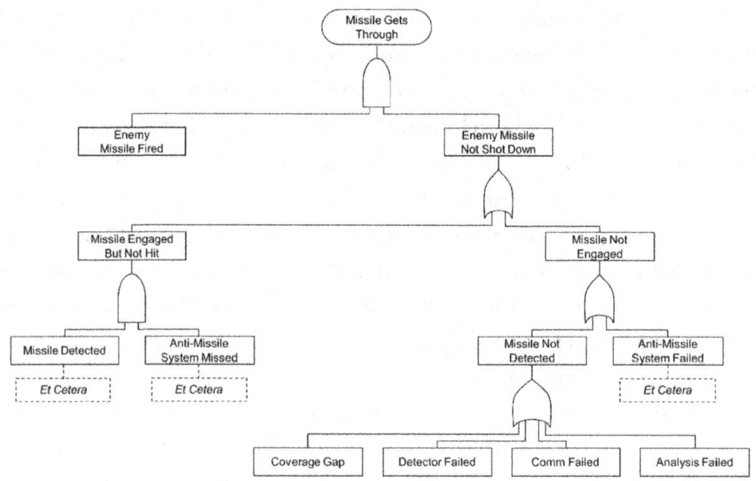

Figure 10. War Tree Specific Example. In this example, the system is National Missile Defense; the "fault" is a successful enemy strike.

The primary difficulty of the war tree is the same as its parent, the fault tree: complexity. In examining a discrete system or limited outcome, like the missile defense example above, it may be possible to account for many if not most of the factors. Even for larger systems, the factors may be discussed, but the effort of doing so—not only in terms of understanding, but of time—is prohibitive. The only question is whether the payoff is worth the difficulty, and in most cases the answer would seem to be no.

In the aggregate, then, the "war tree" is an interesting but unwieldy tool. It might prove useful in certain specific applications where the analysis can be conducted with some amount of ease—e.g., to determine what failures must occur in order to allow a successful bellicosi strike, a successful surprise attack, an unsuccessful air campaign, etc. But the more complex the system being studied, the more complex the war tree and the more difficult the analysis.

26

THE EVOLUTION OF THEORY

We often prefer simple explanations, even when they do not account for every detail; full explanations can strain the limits of our patience and our tolerance for complexity. The constructs we have produced so far, however, are not simple; nor are they fully explanatory.

The first new formulation we examined, that in Chapter 16 of the levels and media of war together, is obviously more complex than treating the levels and media separately. The second, last chapter's war tree, would be a highly complex method of examining war—i.e., of picking it apart to analyze its constituent pieces and their interactions—but it would not qualify as a theory of war. The war tree might be used to explain past events, and may even be useful to predict future outcomes, but would not itself be either explanatory or predictive.

It would be best if we could produce more coherent, useful constructs to aid our understanding of war and conflict. But it remains to be seen whether we can combine more, or more complex, theories to produce anything other than confused jumbles of conflicting ideas. To continue the search, next we will explore the relationship between simplifying an idea and the complexity of interactions; again we will use the "Principles of War" as the basis of the discussion.

In Chapter 22, we established that the interactions between the nine traditional principles of war could easily number in the tens of thousands. If we wish to reduce the number of possible interactions, we may either study the principles with less fidelity or reduce the number of principles themselves.

To some readers, changing the principles, which are supposed to be inviolate, would be military heresy of the first order. Criticize the principles, perhaps; modify them or suggest others, very well; but reduce them? If "this be madness," we hope "yet there is method in't."[1]

One reduced list of principles actually comes from their historian, John I. Alger, who noted in *The Quest for Victory* that four of the principles seem fundamental: mass (or concentration), surprise, objective, and economy of force.[2] This falls in line with Jomini's insistence on the principle of mass, but neglects certain relationships between the principles. For instance, it does not consider any innate relationship between mass and economy of force, nor does it acknowledge that surprise is difficult to achieve without security.[3]

An even more basic list seems possible, but before we begin we should note—once again and for the record—that this type of analysis can be dangerous. It may leave out fine distinctions a more complete model would retain. If we think of the principles as the basic "particles" of warfare, this approach may simplify our understanding in the way that subatomic physics is simplified for most of us in the everyday world. Our everyday experience is with assortments of molecules in shapes we recognize and forms we can hold in our hands; elementary- and high-school level knowledge acquaints us with protons, neutrons, and electrons as the building-blocks of those larger items, but we have no everyday experience with the subatomic particles; higher-level physicists conceptually take the world apart even further, and struggle with gluons, muons, quarks, and the like, but we cannot perceive them nor do most of us really appreciate them except in the abstract. At each level, the increased complexity allows more complete understanding, but most of us simply do not need to study subatomic physics in order to throw a ball or turn a wrench or fire a rifle.

If the resulting model is *too* simple, however, it may not improve our ability to understand the past or predict future outcomes, which is the essence and aim of Theory of Knowledge. By pressing on, therefore, we run two risks: the risk of taking things too far, and the risk of omitting some critical element. Despite the risks, the effort is not intended only to simplify but to understand better. This is the essence of *coup d'oeil*: reducing the mass of data into the few necessary elements of information.

27

ELEMENTS OF WAR

Recall from Chapter 8 that the most basic requirements for a war are an objective, an enemy, and violence. Having gone through Chapter 19's detailed study of the principles of war by considering specific historical examples, we may postulate that, at the most basic level, the necessary elements of *engaging* in a war are: 1) the means of combat (weapons and people), 2) some knowledge of enemy and friendly capabilities, and 3) given the will to engage, a method of controlling the aforementioned combat means in pursuit of the objective. In simpler terms, we must have some way to fight, we must know whom we plan to fight (and when and where), and we must know how to fight.

We can simplify this further. Knowledge of enemy and friendly forces—and here we speak of *knowledge* as described by Theory of Knowledge, not merely data or even information—is critical to determining the direction of and thus controlling the means of combat. The commander cannot direct an attack without such knowledge; it is part and parcel of the command function. Thus we can reduce the larger whole to a basic formulation of only two vital elements, combat means and command of those means.[1]

For our purpose here, "engaging in a war" or a conflict may apply across all the instruments of power, while we reserve "combat" for actions involving the military instrument. Combat applies to fighting a tactical battle all the way up to engaging an enemy with military force at the grand strategic level. (If we were referring to the diplomatic or other instruments, then "opponent" might be a better term than "enemy.") "Combat" applies whether the fight involves troops on the front line or planes in the sky or electrons figuratively bouncing off network firewalls, and whether it refers to an isolated skirmish or an all-out national mobilization.

In addition, "the means of combat" are not contingent upon operational doctrine—either ours or our enemy's. The means of combat exist: they are real, physical entities, whether objects or people.[2] How we, or our enemy, choose to use the means may indeed be contingent upon (or, if you prefer, affected by) doctrine. In other words, doctrine affects command more than it does combat means, especially since doctrine usually lags the development of weapons.[3] In fact, certain classes of weapons, e.g., chemical or biological weapons, may exist but not be supported doctrinally to any great extent.

Consider an enemy poised to strike. We expect them to strike at some center of gravity or another in pursuit of gaining an advantage or securing their objective; with that in mind, we may expect their strike to aim at one of these elements, either our combat means or our command. Even a strike against the general populace—or a segment of the populace as epitomized by the World Trade Center casualties—may be considered an indirect strike against combat means, since troops are raised from and supported the population. If we had neither combat means nor command capability, and our enemy had both, we would be forced to capitulate without the enemy having to strike.[4] It is difficult to imagine any state or nation having no combat means, and equally difficult to imagine any organization as complex as a nation to be completely bereft of command elements, although they may be named something else and be focused on other-than-military functions. So these two elements are extant in any state or nation, and any strike would be directed against them (or some portions of them) to render them less effective. To return the blow—to join actively in the fray—we must possess some combat means and some command capability that we can bring to bear.

In fact, these criteria—combat means and command—are so basic, they apply regardless of the level of war or which media govern the action. At every level of war, from tactical to strategic, in order to attack or defend we need combat means and command over them. No matter what media are involved—land, sea, etc.—we need combat means and command over them. While the *specific* elements for joining a conflict are not universal in application, *in general* some combat means and directive capability must exist, though they may be almost unrecognizable in some conflict paradigms (e.g., irregular warfare). Even guerrilla action requires combat means and command; information warfare, also, requires both elements, although the "combat means" in information warfare may be a fuzzy concept. If we expand the levels to include the grand strategic, still we require combat means and command over them—even though the "combat means" at that level are more properly "conflict means" and include political and economic as well as military instruments of power.

Consider two opposing military forces. Each must have *some* combat means and command, but their specifics may be vastly different. In some cases the combat means themselves will be different, whether in material or training or supply; in some cases the command element will be different, perhaps acting according to different doctrines. In the former, you might think of a heavyweight versus a bantamweight boxer; in the latter, a boxer versus an aikido master.

Drastic imbalances can adversely affect the conduct and perception of a war,[5] and while this is not the general meaning of "asymmetric" with respect to war, it certainly describes a degree of asymmetry. As the old *Joint Publication 1* put it, "Engagements with the enemy may be thought of as 'symmetric' if friendly and enemy forces have similar types of forces and capabilities, or 'asymmetric' if forces and/or capabilities are significantly dissimilar."[6] Consider the Vietnam War. Ho Chi Minh did not have the means or operational knowledge to conduct strategic air operations against the Army of the Republic of Vietnam, nor against US strategic or tactical targets. Nor did he have the means to fight according to the "Army Concept of Land Warfare," the model on which US Army offensive land operations were based in Vietnam. Therefore, Ho Chi Minh had to prosecute one conflict, while for practical purposes the United States joined an entirely different conflict. Above and beyond the fairly rudimentary level of troops exchanging gunfire, those conflicts proved entirely alien to one another.

The two sides defined the conflicts, in objectives and mechanisms and totality, in different terms. That difference was important, but Ho Chi Minh did have *some* means, and knew how to use those means to be effective. Both the Viet Cong and US paradigms, as different as they were from one another, required the control and application of force through some means. Whether that force was napalm or Agent Orange, AK-47s or culture/ideology/psychology, each side developed and used what it thought it needed. The Viet Cong applied their force and we applied ours; apparently they understood the nature of the overall conflict (singular) better than we did, especially as regards the early years of the war. This may be the area in which Gulf War II was most like Vietnam, although we hesitate to compare the two directly because so many politicians and pundits have done so blindly—based on ideology alone—and we do not want to fall into the same trap. In mid-2005, US and Iraqi forces were battling an "insurgent" force made up of foreign and indigenous people; the enemy's approach to the conflict was somewhat like the approach the Viet Cong used, except that it was not primarily directed against the military forces. The Viet Cong worked hard to keep US forces pinned down by hitting bases and supply lines in small military raids, and their attacks planted in the minds of citizens back home the idea of the eventual US pullout. The bellicosi remnants who took up action in Iraq as often targeted civilian and other traditional "noncombatant" targets, and perhaps the clearest similarity between Iraq and Vietnam was that the attacks—and the media coverage of the results—planted the idea in the minds of some citizens that the US should pull out quickly and without regard to the consequences. By mid-2008, a renewed effort by US forces to team up with Iraqi security personnel and local residents to identify and apprehend or eliminate the bellicosi leaders—many of whom were not Iraqis at all—had beaten down but not eliminated the insurgency. Later, a fresh wave of attacks gave the impression that the remaining insurgents simply bided their time (as many commentators predicted they would do) until US forces departed the area before they began another push for power. As much as the conflict waxed and waned, however, both sides had to employ combat means through effective command in order to fight their way toward their objectives.

Asymmetric conflict does not always rise to the level of combat, however, as seen when Mahatma Gandhi contested with the British Empire in his bid for Indian independence. Gandhi did not contend against the British in any open or militaristic way. It may be that Gandhi's lack of obvious combat means itself provided him with the means to his end, in a way so foreign to the British experience that they simply could not adjust. Gandhi's method of fighting without fighting deprived the British of their knowledge of how to fight. Gandhi's paradigm—so alien to British and indeed Western thought—rendered the British ineffective.

This whole approach to "asymmetric warfare" is full of twists and turns, especially where active combat is replaced with more general and less violent conflict. In the end, in any *conflict*—let alone combat—both sides have some combat means, though those means may be different in type and indeed in substance; both sides have some command ability, though the thoughts they think and decisions they make may be different also. What gives one side the advantage over the other may boil down to better comprehension of the situation—environmental, sociopolitical, economic, etc.

To repeat the bottom line of this proposal: to fight a fight, you need something to fight with and you need to know how to fight.

But to examine this construct in greater depth, we must consider one absolutely indispensable element for engaging in a conflict, without which no conflict can take place: Will. The relative will of one combatant versus another has much to do with the extent of a conflict, and certainly whether or not it occurs, and when. Taking the cases of the appeasement of Mussolini in Ethiopia and Hitler in the Sudetenland, insufficient will to risk conflict might be as important to engendering it as the will to fight. Was that not the fear that motivated the US-led coalition to action against Iraq? It took years of Iraqi defiance of UN Security Council resolutions that, as far back as 1991, aimed to destroy their chemical, biological, ballistic missile, and nuclear programs;[7] plus continuing attacks on US personnel enforcing the "no fly" zones; plus the unprovoked action of the bellicosi; to elevate our will to fight—and underlying the will was the fear of what might happen if another tyrant continued to be appeased.

Consider, with respect to will, the "official" Joint Chiefs of Staff definition of "command"—which, when referring to orders, specifies "the will of the commander expressed for the purpose of bringing about a particular action,"[8] making it no surprise that battle orders begin with the "commander's intent." "Will" is a vital component of "command" (and, indeed, a vital component of "combat means" if they are taken to include troops and their morale and readiness). It is interesting to note, in this regard, that although the German army developed a reputation for fierceness, Hitler reportedly "showed anxiety about the 'will to war' of the German soldier once the exhilarating effect of the easy conquest of Poland wore off."[9]

Intrinsic to the question of will is having enough will to *win*—having enough to *risk* conflict, enough simply to begin to fight, will not always carry the day. The US in World War II had enough will to vaporize nearly all the industrial and urban centers of Germany and Japan, civilian populations be hanged,[10] but only a generation later, in Vietnam, lacked the will to do the same during "Rolling Thunder." Without the will to fight and win, who would expect to win?

This is the fundamental question that ebbs and flows in the tidewater of US policy: whether we as a nation have the will to finish what we start. In the ongoing Terror War, our bellicose enemies bet that we do not. Either we will prove them wrong or we will confirm their assessment; if the latter, it will be to our unending peril. This may be part of the reason that the new principle of "legitimacy," which includes cultivating and maintaining the will to win, was added to Joint doctrine.

One example of the importance of the will to fight is actually two: the dual battles of Hakata Bay. The first battle, in 1274, saw Kublai Khan's 25,000 Koreans and Mongols initially beat the Japanese defenders back from the coast, but the invaders pulled back to their ships when a storm struck. They eventually returned home and busied themselves with settling domestic matters in occupied China. The hiatus gave the Japanese both time to prepare and a glimmer of hope that the Mongols would not stomach a real fight. The second battle occurred in 1281, by which time the Japanese had built a wall to restrict the Mongols' access to the coast and developed new tactics of close-order ship-to-ship battle. The Japanese defenders held off the 150,000 Mongol invaders for 53 days, at which time a typhoon (in Japanese: "kamikaze," or "divine wind") decimated the Mongol fleet.[11] It may be that defenders naturally manifest a stronger will to fight than invaders, or that the Mongols had greater will to fight than their conscript Chinese and Korean partners, or that the Mongol invaders had enough will to fight the Japanese but not enough to fight nature itself; whatever the case, the Japanese won, and that 1281 victory solidified in Japanese warrior culture the supremacy of the sword, an idea that would be held in esteem up to and through World War Two.[12]

The US fighting will, has been somewhat different. While the fighting spirit may have developed out of the pioneering spirit, whereby we fought fiercely and often with no quarter asked or given, our will to fight has never been constant. From our earliest days, after almost every conflict we divested ourselves of much of our military might. In the

1960s our national will to fight fell to its nadir; and in the waning decades of the 20th Century, despite maintaining a large and effective fighting force, we fell back from operations in Lebanon and Somalia and gave would-be enemies the clear signal that we truly did not have the stomach for a real fight. We withdrew from Iraq in 2011—only to return within a few years as the threat from the Daesh bellicosi grew—but whether we stay the difficult course ahead there, in Afghanistan, and elsewhere, remains to be seen.

28

THE ELEMENTS AND THE PRINCIPLES

If this construct of combat means and command of those means describes war with any accuracy, then it should be possible to fit the classical "Principles of War" into these categories. If we do so, what form they will take?

As pointed out in Chapter 19, mass and economy of force seem linked intrinsically, as the latter often explains failures (and indeed includes the definition) of the former. Both principles refer to the availability of "combat power," and, if you will pardon the continued use of physics terminology, together they may be referred to as "potential." Potential energy is stored energy, energy of position: a boulder on the side of a mountain has potential energy, but its energy is only released when it falls and strikes the ground. So too combat power—mass—remains potential until we release it. (This may be a useful analogy, but like all analogies we should not make too much of it.)

Likewise, the principle of the offensive is intrinsically linked with maneuver: combat means which cannot move cannot go on the offensive. While it is true that some maneuvers are not inherently offensive, such as feints, no offensive action can be undertaken without employing maneuver. Recall that in Chapter 19 we emphasized quickness and discussed that the principle of maneuver itself may be better rendered as "speed."

Putting these two ideas together—applying the means as opposed to just possessing them—is essential to prosecuting any battle. The ability to move with nothing to move is as useful, or useless, as tons of materiel that cannot move to where they are needed. A C-17 with no cargo to transport is as useful to the Combatant Commander as the cargo with no means of getting it to the troops. Applying combat means—changing their potential into kinetic energy—is more important to winning a battle than their availability. We may think of this basic need as combat means *in motion*, and may call this construct, for the purpose of this exercise, "momentum." The term "momentum" may appeal to purists, and perhaps might even have appealed to Jomini, since mathematically momentum equals mass times velocity. Napoleon may have been the first to recognize the importance of momentum in martial terms.

For industrialized nations, the concept of "momentum" may include the nation's industrial capacity, i.e., the ability of a nation to maintain enough production to sustain a war effort. US industrial might, for instance, was a key factor in our victories in both theaters of World War Two. This would not apply in the same way to the case of insurgencies, and maybe not in limited war where heavy fighting is the exception rather than the rule, but it certainly figures in a nation's overall ability to wage war. In the current global marketplace, however, this should probably include a nation's (or group's) ability to purchase weapons—or to obtain them by illicit means. Thus the continued emphasis on and effort toward nonproliferation activities: tracking weapons materials and precursors; limiting transfers of key weapon technologies; intercepting shipments of weapons components and related equipment; etc.

Moving on, three of the classical principles—unity of command, objective, and simplicity—together may be considered under the well-known name of "command and control," or C^2, by which the commander focuses forces on the objective by means of simple, direct orders. While "command" and "control" sometimes refer to different aspects of the commander's role—the issuance of orders (directives) and the maintenance of order (discipline)—they are practically synonymous with one another.[1] Joint Doctrine lays out the tenets of good C^2 as:

Clearly Defined Authorities, Roles, and Relationships
Mission Command
Information Management and Knowledge Sharing
Communication
Timely Decision Making

Coordination Mechanisms
Battle Rhythm Discipline
Responsive, Dependable, and Interoperable Support Systems
Situational Awareness
Mutual Trust.[2]

Most of these apply to both the command and control functions: e.g., clearly defined relationships are vital to issuing orders and maintaining discipline; communication and decision making each feed into command as well as control; and so forth.

Command and control cannot be effective, however, without knowledge of the enemy, which is why additional letters are often appended to C^2 for intelligence (I), surveillance (S), and reconnaissance (R). Indeed, the construct grew to enormous proportions at the end of the last century. Every couple of years, it seemed that someone found another inadequacy in the notion, as letter after letter was added to the acronym. C^2 became C^3, C^3 became C^3I, and C^3I became C^4I, all rather haphazardly as communications, computers, and intelligence were added. The additions of surveillance and reconnaissance pushed C^4I into C^4ISR or other combinations (e.g., C^2ISR, as in the one-time Air Force C2ISR Center at Langley Air Force Base, Virginia[3]). But is the C^2I/C^3I/C^4Ietc. relationship something of a chimera?

C^2 seems preferable over C^3 for the simple reason that without even rudimentary communications there can be no C^2; whether verbal or semaphore or telegraph or satellite, *some* means of communications is essential. Therefore, the extra C may be assumed. In addition, "command" and "control" refer to actions. "Communications" and "computers" refer to capabilities and tools, while "intelligence," "surveillance," and "reconnaissance" refer to inputs used to make command decisions. The last few are the results of actions, yes, but they are generally not actions taken directly by the combatant commander; other commanders take the subject actions on the combatant commander's behalf. Indeed, the combatant commander is more concerned with the *results* than the actions taken to achieve them, so long as the actions were lawful. Take, for example, intelligence regarding the scope of an anticipated attack: The contents and veracity of the information itself is more important than whether it was gained from a captured communiqué, an enemy defector, or a chance observation of the enemy's movements. Or consider satellite imagery: The timeliness and usefulness of the end product, rather than the details of its production, matter more to the commander in the field. (Thus, to take a famous example, the information that confirmed the location of Osama bin Laden's hideout in Pakistan was more important to Seal Team Six than how that information was gathered and verified.) So, the absence of these additional inputs may make command more difficult, but command still is the central, essential element.

Considering these relationships, think about the link between the final two classical principles, surprise and security (with security's cousin, deception). Though these deal explicitly with hiding our abilities and intentions from the enemy, they also imply a corollary: that we know something of the enemy's abilities and intentions. As part of the commander's situational awareness and as resultants of the commander's decisions, we may include these principles under the overall heading of "command and control," which we may again simplify into "command."

Recall now the earlier discussion of the national or individual *will* to fight. Should we include in the concept of command and control the overall willingness to wage war? It seems a case could be made both ways. We may consider the will of the civil and military leadership to wage war or prosecute battles as a necessary precursor to the actual issuance of orders, in which case the two would be joined together. Or we may treat the will to fight as something distinct and separate from the ability to fight, and include command and control on the side of ability rather than will.

It is important to note that, for most peace-loving peoples, the will to fight is different from a *desire* to fight. A bully has the desire to fight, and therefore the will. Someone who has much to gain from fighting may develop the desire and thence the will to fight. Others without the desire may avoid fighting until pushed too far, at which point the will to fight may manifest itself. But without the will to fight, and especially the will to win, any efforts will be at most half-hearted. Therefore, will appears to be an integral part of command and control, with the strength of command being at least proportional to the will to fight.

Therefore, the nine classical principles can combine and recombine into the two postulated elements: applied combat means (or, Momentum) and Command. These two elements combine features of the Principles of War and seem indispensable to military action on practically any scale. Momentum includes personnel, weaponry, transport, and logistical support. Command includes what we know—

about ourselves, about our enemy, about our objective—and how we apply what we know.

"Command" also includes the nature and tendencies of the commander. This seems odd at first, almost as if we were using a word in its own definition. If the commander is the one evaluating the situation and making the decisions, can he be considered as an input? General Patton apparently thought so. Prior to facing General Erwin Rommel in the African desert, Patton studied what the German leader had written about warfare—and Rommel himself studied the theories of Britons J.F.C. Fuller and B.H. Liddell-Hart.

The qualities of the commander (especially the enemy commander, if they are known) can influence how the commander approaches the battle, and therefore the outcome of the exchange. Is the commander incisive? Decisive? Bold? Reckless? As Sun Tzu wrote,

> There are five qualities which are dangerous in the character of a general. If he is reckless, he can be killed; if cowardly, captured; if quick-tempered you can make a fool of him; if he has too delicate a sense of honour you can calumniate him; if he is of a compassionate nature you can harass him.[4]

We will not go into specific examples here, but you may wish to consider, for instance, what we knew of Saddam Hussein that influenced our planning and execution of Gulf War Two.

In addition, knowledge of the situation and the enemy may dictate different command approaches at different times. Is there a time to be cautious instead of reckless? Caution would have well served Southern forces at Gettysburg. Is there a time to be direct instead of indirect? Liddell-Hart apparently did not think so, or at least preferred the flexibility of the indirect approach. Is it necessary for the commander to change command style to fit the situation? The prospect may be nearly impossible for some—like asking the leopard to don the tiger's stripes for a single hunt. Or must the commander continue along a path dictated by his personality? If so, does the enemy know his personality—and therefore know what he is likely to do? It would seem that many great generals—Hannibal, Napoleon, Rommel—were eventually defeated precisely because their enemies studied them so well.

As Sun Tzu advised, "Know the enemy and know yourself."[5]

As we attempt to understand the relationships between these elements of war, and how they work together in the practice of warfighting, we should consider the effect—the sometimes overwhelming effect—of circumstance on the outcome of a battle. Note that by circumstance we do not mean chance. The effect of chance and the fog of war may be significant, but only rarely can it be controlled or even altered. Chance can only be reacted to. Rather, in this case circumstance refers to the conditions—the weather, the topography of the terrain, etc.—that may be exploited to one side's advantage. Sun Tzu wrote as extensively on this subject as on any other, in his chapter on "The Nine Grounds."[6]

This, as much as anything else, is a case in which *coup d'oeil* provides a prepared commander with insight an unprepared commander may lack; or perhaps provides a gifted commander with opportunities which a merely adequate commander would miss. The prepared commander sifts through the given chaos of the situation, quickly assessing the prevailing circumstances and how to take advantage of them (along the lines of Colonel Kratman's principle of "Shape," from chapter 20). Is there high ground for observation and from which to pound the enemy with withering fire? Are there points where the enemy may be trapped? Is a storm coming, and can we operate in it better than the enemy? Such considerations might effectively change one side's Momentum (in the construct presented in this chapter) although superior Command would be necessary to exploit them.

It may not be possible to overemphasize the importance of preparation. Not only personal preparation through study and training, but preparation of the larger enterprise through careful, thorough planning. With that in mind, one aspect of this Command and Momentum construct that may determine if it has any utility is whether —and if so, how—the ideas fit into planning military operations.

In the planning process, combatant commanders project strategy downward from national aims, ultimately all the way to tactics—using the operational level to broaden tactics and also to constrain strategy to fit the commander's needs. Ideally, advance planning takes place in the relative leisure before a crisis erupts, and it is good to remember that crises generally do not erupt spontaneously: even if the causes and their warning signs are unforeseen, undiscovered, and unexamined, the causes exist. Crises begin, continue, and end; so do campaign operations;

which calls for a continual process in which plans are developed and modified in response to intelligence, alliances, and other factors of the world condition. It should not have surprised anyone, for instance, that nascent plans for conflict in and with Iraq existed and were updated before the Gulf Wars—at least, not anyone familiar with what was then called "deliberate planning." What was somewhat surprising is that those contingency plans were later derided as if they proved that an invasion was a foregone conclusion.[7]

What would be truly shocking, and indeed close to dereliction of duty, would be a refusal on the part of the military to plan for eventualities. "Crisis action planning," on the other hand, was—as the old name implied—done in response to the crisis. The US Department of Defense recently streamlined its approach to planning, and removed the "deliberate" and "crisis action" terms from the DoD lexicon in favor of treating contingency plans as "branches and sequels" to larger and ongoing campaign plans;[8] with apologies to the powers-that-be in the Joint Doctrine community, and perhaps as an indication of my being an old dog that refuses to learn a new trick, I still prefer to think in terms of "deliberate" and "crisis action" planning. But, regardless of the naming convention, military leaders must still plan carefully and adapt their plans to changing situations. Doing so may consist of updating an existing plan to fit the specific scope of the crisis, or crafting a new plan on the spur of the moment. If the former was the case for the campaign in Iraq, the latter was the case with the first campaign of the Terror War: defeat of the Taliban and the liberation of Afghanistan in Operation Enduring Freedom. We will not delve into the intricacies of the two planning processes; suffice it to say that the advantage to linking the two is that contingency planning makes crisis planning easier and ensures that, should a crisis arise, some amount of well-thought-out planning will already have been done.

Do the two elements of Command and Momentum simplify or confuse the planning process? So long as we do not limit our viewpoints and so miss possible inputs to our plans, they should at least not hinder it. They will not directly help develop possible courses of action except where they give us insight into what advantage, if any, we hold over our adversary and how our advantage may be improved. They are better suited to providing touchstones for evaluating and selecting courses of action, but do not, in and of themselves, provide the level of detail to produce a near-complete picture of the order of battle. In the end, unfortunately, these two concepts do not lend themselves easily to assisting with campaign planning, except perhaps to remind planners to consider which courses of action would maximize our Command and Momentum while minimizing our enemies'. We will consider in later chapters whether that condition is a good predictor of eventual success.

29

THEORY VERSUS THEORY

When we presented the two elements above, we claimed without elaborating that they bear some amount of universality: that they apply regardless of level or medium of war. In other words: In every case, from tactical to strategic, from land to space, we must have some Momentum (combat means) and Command over them in order to fight and win. Is this really true?

Consider the simplest level of tactical conflict: one on one. Whether the participants use bare hands, fists clenched in boxing gloves, knives, dueling pistols, or broken bottles, both will apply those chosen means of combat by using the raw function of command: they will assess each other's strengths and weaknesses, and try to land effective blows while avoiding their opponent's strikes. (This is especially true when the fight is asymmetric, e.g., sword vs. knife.)

Each step upward, from soldier to squad to platoon, we see the same pattern. The two sides may be armed similarly or differently, but still they use whatever combat means are available. At some point we move from "fight" to "battle," and we see more structured organization under a system of leadership. As we move upward, usually the combat means become more destructive—machine guns, artillery, armor, gravity bombs, precision guided munitions, missiles—and the command functions more complex, but still the pattern holds.

At higher levels, up to strategic and beyond, the scope of battle grows. Here, coordinated attempts must be made to evaluate the enemy's capabilities and intentions, and the overall combat power at the commander's disposal becomes greater and greater, but the ultimate aim still is to land decisive blows that will achieve the battle objective. Not necessarily *the* decisive blow, as so favored by Jomini, although the "American way" of war or battle has tended that way in its pursuit of annihilation of the enemy's armed forces.[1]

As an aside, you may have noticed that much of this text has purposefully neglected, if not come close to denigrating, the operational level of war. As discussed in Chapter 12 and elsewhere, it is difficult to go through an analysis such as this without wondering if the "operational" level is an unnecessary construct. Shoehorned into place—if not driven into place like a wedge—it seems to shift the tactical level into a smaller space than it used to occupy, almost supplanting the tactical level in some cases. Much effort is put into understanding the operational level of war, when as a concept it might be something entirely different from the "operational art." Still, in that nebulous area where tactics and strategy overlap, both combat means and the ability to command them must be present and functioning.

Back to the subject at hand. What about "lower" levels, such as irregular warfare? Does the Momentum and Command construct apply here as well? Yes. Regardless of what type of operation is being planned and carried out, some type of military power is still being controlled and brought to bear, even if it is not strictly combat power in the traditional sense: as historian Charles Townshend observed, "High-intensity war was classically Clausewitzian: speed of strategic manoeuvre, concentration of force, the culminating point, decisive battle. In low-intensity war, these qualities are reversed: fluid positions, diffusion of force, and protracted psychological attrition are the means. . . ."[2] In effect, considering that irregular warfare can serve national- and even international-level interests, it might be said that instead of being sub-tactical irregular warfare operations are actually grand strategy practically applied.

With respect to the media of war, these elements also work. For each medium in which battles may be fought, we can identify combat means at a commander's disposal. (The combat means for the medium of space are still hypothetical instead of real, fielded forces, but at one time such combat means as tanks, airplanes, and submarines were also the stuff of fancy.) For every medium, command decisions influence (or, for space, will influence) how combat momentum is applied against enemy forces.

As discussed in Chapter 15, network-dependent militaries today wrestle with how to categorize and understand information warfare, and whether it represents a medium unto itself. We may concede the importance of understanding the electronic realm in military terms, but it seems the migration of conflict into "cyberspace" is a contributing factor to the dying military art. As much as those who cry only for peace without cost would protest the notion, wars throughout history have given warriors the opportunity for glory—but how much glory may there be in directing a computer attack? The art of war, then, may languish until translated into other realms where men live. Space, therefore, as a place where humans are likely to live, lends itself more readily than the information realm to the traditional military art. Regardless, information battle, like any other battle, involves some combat means—be it malicious logic, electromagnetic pulse, or something else—applied as determined by a command function of some sort.

Recall now the combined levels/media model presented in Chapter 16 (Figures 3 and 4). Does it support this dual-element idea? Given that combining the levels and media did not change their natures but only arranged them so they were interdependent—and that we demonstrated above that Command and Momentum apply in each medium and at each level separately—then they must apply here also. In fact, referring back to Figure 4, each of the operations identified that span two or more media must, as with their single (or intra-) medium cousins, involve the command of applied combat means.

This Command and Momentum construct also holds true when considered along with other theories. The "indirect approach," being a prominent vision of maneuver warfare, requires combat forces in motion (i.e., Momentum) controlled by a commander who can decide between different avenues of approach and select the best with which to freeze the enemy in place. Colonel Warden's "Five Rings" theory assumes that the command element will be able to analyze the enemy's systems to determine the critical nodes, then direct combat power against those nodes: Command and Momentum must work together for the "Five Rings" model to have any usefulness.

In the end, this simple construct—two elements, Momentum and Command—will never be comprehensive. It may be foolish to try to develop a comprehensive theory; even Clausewitz and Jomini were unable to do so. If we tried to postulate all the elements a truly comprehensive theory should include, our list likely would be too long to consider the possible relationships and interactions.[3] Even this abbreviated notion may still be *useful*, however. If it holds true with respect to some of the concepts by which we understand war and warfare—which is what the levels, media, and other theories are—how does it fare when tested against reality? Does it also hold true across the range of military history?

If we test it against historical encounters, battles, and campaigns, and against hypothetical future encounters, we may find it an adequate explainer or predictor. Recalling the basics of Theory of Knowledge as presented in Chapters 4 and 5, and especially the key point that all knowledge is probable only, we will *not* find a perfect predictor. But we may find one that meets our needs.

To test the construct, we must examine it in light of other theories and in light of history. When we test it, however, we must take care not to yield to the temptation to select only examples which prove the thesis.

Testing even this simplified theory against experience will also prove arduous because of a weakness admitted earlier: the dearth of my own historical knowledge, given that I am (at best) an amateur student of history and very much a product of this modern era. Therefore, we will not attempt an exhaustive study of military history; however, we will try to frame this theory within the context of history. While doing so we recognize that, despite our best efforts, we may merely be tilting at a theoretical windmill.

30

EXAMINING THE "INTERACTIONS"

The two elements, Command and Momentum, both operate on each side of the front. That is, each side of a battle presumably has some combat means and some mechanism for command. Based on this, at first it seems possible to examine them as four separate elements, i.e., friendly momentum, friendly command, enemy momentum, and enemy command. Taken at only two levels, weak and strong, this yields sixteen interactions.

This analysis, however, produces confusion for two reasons. First, a battle between a "friendly" state high in both elements and an "enemy" state low in, say, command, should produce the same analytical results as one between a "friendly" state low in command and an "enemy" state high in both elements: the notations are the same, and therefore the affiliation should be irrelevant. I went around and around that axle for a long time, trying to compare cases where both combatants were thoroughly weak in command and momentum, or where one combatant was physically or materially strong but weak in command while the opponent was weak in both, and so forth. It proved fruitless—or I proved unequal to the task.

Second, determining the "friendly" and "enemy" elements in a battle between two foreign powers may be problematic for vicarious observers in a third state. Since we are concerned primarily with one force versus another, and less with which one is actually on our side, the assignment of "friendly" and "enemy" to the cases is merely a convenience for the sake of understanding. Therefore, it can be neglected when we examine the cases in detail.

A far better analysis seems to manifest when we assume that *absolute* strength or weakness in the two elements does not matter as much as *relative* strength or weakness. That is, whatever ranking we might assign to our own momentum does not matter as much as how our momentum compares to our opponent's, and who holds the advantage. The search in warfare is for an advantage—from Sun Tzu's "calculations" and "estimates" to Jomini's insistence on numbers to Boyd's OODA loop itself. We might even go so far as to say that the search in *life* is for advantage: In commerce we seek to control the market; in diplomacy, to significantly influence (if not outright control) relations with and between states; etc. In social settings we seek the upper hand so that we might be more in control of the situation, and we cling to whatever advantage we perceive that we have, be it wealth or power or intellect or attractiveness. This line of inquiry would make an interesting sociological study, but it is beyond the scope of this text.

The pressing question when considering these proposed elements of war is what advantages, if any, guarantee or at least predict military victory. Now and always, according to Theory of Knowledge, how do we sift through the "chaos of the given" to reach the key information we need to arrive at knowledge, i.e., at reliable predictors?

We cannot consider every factor, unfortunately. Recall the brief discussion earlier of the circumstances of battle. Part of the "chaos" with which a commander is faced includes environmental factors: the ground/terrain, routes of approach and escape, weather, etc. These affect command decisions and the operability of combat equipment, but they are not under the commander's control; these, more than almost anything else, guarantee that the play of chance will have some influence on the outcome. Moreover, they affect both sides of the conflict, though not to the same degree; e.g., both sides may covet but usually cannot occupy the "high ground" at the same time, which speaks to how the prepared commander can use such factors advantageously. However, since no theory can hope to encompass all of these factors, we are reluctantly forced to neglect them in this analysis.

So, considering only the elements derived earlier—Command and Momentum—and focusing on which side holds the advantage, the resulting "interactions" are shown in Table Four. We will examine each of these interactions beginning in the next chapter.

Table 4. Interactions of Advantages.

CASE	MOMENTUM	COMMAND	SHORTHAND
1	Even	Even	Parity
2	One side with advantage	Even	Single Advantage: M
3	Even	One side with advantage	Single Advantage: C
4	One side with advantage	Other side with advantage	Split (or Mixed) Advantage
5	One side with advantage	Same side with advantage	Double Advantage

We will examine each case from Table Four using the same pattern: identify the case; consider a slate of historical examples that seem to fit it as well as some that may contradict it (in order to avoid the error of selecting only those examples that support the premise); and offer future scenarios, if possible. If appropriate, we will offer some conclusions about how well the elements apply or the need to modify this description.

Before we examine the individual cases, we may make some generalizations about the types of examples that represent them. Each case weighs the advantages in command and momentum held by one force over another, comparing the forces' preparedness and abilities with respect to the two elements. The factors making up each element generate clues as to how they may change.

Momentum may be affected by the availability of weapons, their sophistication, the adequacy of training, or the ability to transport the weapons to the point of conflict. Strong momentum requires well-trained, well-armed, well-supplied troops. Weak momentum may be caused by primitive or inferior weapons, even with comparatively good transportation and logistical support. The same result may be the product of another equation: excellent weapons that cannot be brought to bear on the enemy may also contribute to poor overall momentum.

Command may be affected by the abilities of the commanders (including subordinate commanders), the availability and reliability of communication, knowledge of the enemy's abilities and intentions, and the clarity of orders and instructions. Strength in this area implies that the commanders are able, knowledgeable about their situation and their enemy, and can communicate to and with their forces. Weak command, however, may be due to inept commanders (even with unimpaired communications), limited communications that hinder otherwise capable commanders, incorrect interpretation or use of information about the enemy, or enemy espionage and reconnaissance that nullifies any information gathered.

As mentioned several times now, it is important to remember that each element actually operates along a continuum from poor to good. Rarely will a combatant be completely in possession of or devoid of one or the other element; that is one reason why the relative strength of the two combatants is more useful than some absolute measure. Even so, anyone may debate why we include a particular historical example in one case as opposed to another. As such, we will not reach any ultimate truth in this analysis; but hopefully we can shed light on the dark spots of military prediction.

31

CASE 1—PARITY

In this case, both sides of a conflict have roughly the same Command and Momentum. Whether the two combatants are equally poor (e.g., two small states which might suffer in comparison with larger nations but are equivalent with each other) or equally strong, it can only be said that, on the surface, neither has an obvious, decisive military advantage. Whether the two commanders are equally competent or equally inept does not matter. By virtue of assigning them to this case, we assume that neither has better insight into the situation, or more sophisticated communications capability, or more effective intelligence gathering. In the main, neither side seems to have an advantage.

The ultimate "null" case, in which both combatants are thoroughly weak (e.g., two fledgling communities at the dawn of civilization), serves primarily to illustrate the danger of a complete lack of preparation—a lack that carries over into all levels and types of war. If the combatants are strong in some fashion but their strengths only match each other, these may also turn out to be null cases regardless of where they are on the industrial scale (recall the types of war presented in Table One).

The long-running and, in the long run, inconclusive, war(s) between Iran and Iraq in the 1980s might exemplify another parity case. Each combatant had significant, though not overwhelming, military might—Iraq in materiel, Iran in personnel—but neither demonstrated outstanding command and control. The fighting was fierce, but the result was a near stalemate.

We might also consider the Cold War as a parity case, but only if we ignore the ability to sustain the *economic* contest. From a strictly military viewpoint, especially as the so-called "gaps" (e.g., the "missile gap") turned out to be smaller than advertised, the approach to nuclear preparedness was a search for parity bolstered by the 1960s-era McNamara-inspired doctrine of Mutually Assured Destruction. Each step—atom bomb, hydrogen bomb, intercontinental ballistic missile—was only partially a step toward superiority, as the destructive potential of nuclear weapons cowed even those most determined to wield them. Not to limit this discussion to the momentum afforded by nuclear weapons, consider the command element: The US may have enjoyed an advantage in command of our nuclear forces, but the country intentionally enforced constraints against using them, evening-out the two sides. The US's refusal to develop and field the neutron bomb, for instance, indicated that—at least on the policy level—parity between the superpowers was desirable.

It seems, then, that we can discard this case, not because it does not exist in history but because it is not helpful in preparing for victory. Maintaining parity is costly enough, but a war in which we are on par with the enemy will be even more costly because it is likely to be a long, bloody, drawn-out affair; recall, also, that Sun Tzu warned against conducting protracted wars.[1] Better to find and press an advantage. In the end, since all sides in a conflict seek an advantage, and where possible a decisive or even asymmetric advantage, we are better off examining those advantageous cases.

32

CASE 2—SINGLE ADVANTAGE: MOMENTUM

The case of advantage in momentum is not as easy to categorize as it may seem. It may be convenient to start with the obvious—numerical superiority or superior weaponry—and work our way to more difficult examples.

Many classical battles follow the pattern of momentum carrying the day. At Thermopylae, the gallant Spartans were eventually destroyed by their foe even though their betrayal through espionage would fall into the "command" case. Caesar's conquest of Gaul was predominantly a victory of superior Roman troops against less organized and less well-trained irregulars (the brief sally by Vercingetorix providing a telling counterpoint). Even the well-known "Pyhrric victory" takes its name from the commander in, and is now applied to, battles in which momentum won the field despite the effects on the victors. Thus it is easy to see why Jomini would seize upon mass as his critical principle of war.

Moving from human to technological advantage in momentum, we might consider various examples of Imperialism. One is the westward push of the US, especially cavalry engagements against the natives. Before the US Civil War, the Maxim and Gatling guns were not yet invented, but good rifles were, so pure combat power was somewhat limited but still superior to the bow. Once the Indians obtained rifles, a short-lived technological parity existed until rapid-fire weapons came on the scene, at which point the balance shifted again to the US cavalry. Command was fairly even, except perhaps where it dealt with logistics, intelligence, and communications. Settlers started out with limited supplies and were largely ignorant of the cultures they encountered, ignorant of their strengths and perhaps arrogant because of it; while the cavalry did not have immediate access to technological tools such as the telegraph.

When command was equal but weaponry superior, was the outcome predictable? Many times it was, such as in the high-momentum versus low-momentum colonial campaigns in which African tribes threw wave after wave of warriors against rapid-fire weapons and rifles equipped with dum-dum bullets.[1] Memories of these massacres may be one reason the Boers took to the "commando" trail instead of engaging the better-equipped British in head-to-head battles. Even the Italian army, not known for producing great commanders since Roman days, enjoyed victory over technologically inferior tribes in Africa. This type of momentum advantage is frequently a determining factor in wars between technologically advanced peoples, also. As John Keegan writes, "when men of equal worth fight on unequal terms, the side with the better weapons wins. . . . It was to be the lesson of both Germany's world wars in the twentieth century, when their military caste's contempt for their enemies' superiority in the *Materialschlacht*—the battle of attrition —ultimately availed their soldiers' courage not at all."[2]

In the interest of fully examining this case, can we identify a contrary example?

In December 1944, the combined Allied momentum was pushing the German army back to its homeland. But in the Battle of the Bulge, the Germans overcame the Allies, for a short time, until the sheer force of Allied strength kept them from breaking out entirely. This illustrates the property of momentum as being more than simply weapons and men: it is weapons and men in motion, applied against the enemy. Measured solely in material terms, the Allies held a great advantage, yet the Germans applied their material to win a short-lived victory. This example also shows the difficulty of fitting historical events into clean categories: it might fit better as a "split" case than a case of advantage in momentum. The Battle of the Bulge might, in addition, demonstrate that the availability of superior force counts more in long-term engagements than in short battles, though the US experience in Vietnam would seem to cast doubt on that claim.

The more examples we find where greater battlefield momentum led to success, are we in danger of drawing the conclusion that advantage in the material realm equals victory? Is this, after all, only a restatement of the Jominian emphasis on mass? It may indeed be so, but recall that momentum does not have to mean overwhelming force. Physically, a small mass moving very rapidly can have the same momentum as a large mass moving slowly; in warfare, a small unit striking ultra-fast may disrupt the enemy as much as a large contingent rolling in more slowly. Was this not the very approach used consistently to good success by Lawrence of Arabia?[3] Was it not used also against the US by the Somali warlords, and by the nineteen bellicosi on 11 September? Furthermore, the idea of small units or even individuals working together to generate a large amount of combat power is the very substance of revolution. We may use this understanding to characterize our approach to irregular warfare.

Back to "conventional" or "traditional" warfare, Case 2 also seems to describe sieges and naval or other blockades against an adversary. Sieges, from Masada to the Alamo, involve prolonged application of force even without active fighting; patient strangling of the fortress can deliver it with the loss only of time, while siegecraft, sappers, and guts may also win out for fortunate attackers. Other historical cases illustrate this, such as the Allied blockades that strangled Germany in the World Wars and the US "quarantine" of Cuba in 1962. In each case, applying sufficient naval strength and national resolve led to the desired results.

Conversely, when the attacker is unable to maintain momentum—e.g., the blockade is not complete—the attacker finds it more difficult to win. Indeed, this might actually be a shift of momentum, however brief, to the defense: the attacker's combat power is static, while the defender is able to *move* combat power (in this case, food and other supplies) to where they are needed. Economic sanctions that turn out to be porous or inconsistent (we may think of Iraq and the "oil for food" debacle) may qualify as noncombat examples.

In another way, this case may illustrate the "secret battles" of the Cold War between nuclear missile submarines and attack submarines.[4] Here, the superior maneuverability of fast attack submarines gave them the advantage in momentum, at least on the tactical level; i.e., a "boomer" has high overall momentum because of its immense arsenal of ballistic missiles, but they cannot be used against another submarine. This also illustrates how difficult it is to fit specific instances into these five cases: If the momentum of both were considered roughly equivalent, the primary focal point of these engagements would be the search for useful intelligence and the attempt to overcome the inherently poor tactical conditions of the underwater arena; this would effectively yield Case 1 (parity), but involve the effort to move to Case 3 (advantage in command).

As mentioned in the last chapter, if we shift our focus away from the strictly military instrument of power we may also consider the economics of the Cold War. Militarily, both the US and the Soviet Union could boast good command: the flexibility of distributed control may have given US forces an advantage, while Soviet success at obtaining US secrets eventually evened the overall score. Despite this, the momentum element weighed heavily on the side of the US, but it was momentum of economy, not necessarily weaponry. After all, the collapse of the Soviet economy—from trying to maintain its defense industry and infrastructure—was a primary factor in the dissolution of the Soviet Union.

The components of these instruments of power are not static, of course. The economic instrument suffers through recessions, for instance; its flexibility may be a key to its strength, although the interdependence of the global marketplace may affect some nations more than others. Considering that US military momentum has suffered (since even before 11 September 2001) from drawdowns, base closures, defense industry mergers, and increased operations tempo, we may hope that continued economic recovery will be strong enough to make the most recent recession a comparatively short-lived downturn. But, in the aggregate, these constantly fluctuating conditions present special problems for contingency planners and would-be grand strategists.

The ultimate possibility to consider in Case 2 is the waging of "total war" on one side versus "limited war" on the other side, as seen in the Vietnam War. For instance, despite the end results of the Tet offensive and the ultimate fall of Saigon, it cannot strictly be said that US forces were low on momentum. Our fielded armies did not lose a single conventional engagement—including Tet, which was more of a surprise and a shock than anything else—but our technology and strength did not prevail in the unconventional war. We had power, but were unable to apply that power effectively—though the strategic bombing proponents who tout the effectiveness of "Linebacker I" and "Linebacker II" in forcing Hanoi back to the negotiating table might disagree. As discussed in Chapter 27, the results were inconclusive because the two sides were waging essentially two different wars.

33

CASE 3—SINGLE ADVANTAGE: COMMAND

It is tempting to point to storied commanders—Alexander, Hannibal, Caesar; Napoleon, Nelson, Patton—as the catalysts for advantages in command, and there is no doubt some truth in that. But command means more than just who is at the helm. The skilled general facing the novice holds a great advantage, but not an insurmountable one if the novice perceives the situation more clearly and acts accordingly. As an example, we already mentioned Vercingetorix, the Celtic leader who temporarily held Caesar at bay in Gaul; it may be said that great generalship was not the key to Rome's eventual victory.

We might search first for ancient campaigns and battles in which knowledge of the enemy or the superior use of forces proved to be the decisive factor. In our search, we may choose not to consider the Israelites taking Jericho under Joshua's leadership, because the direct intervention of God would make any preparations (or lack thereof) moot; on the other hand, if we consider that the seven days' march around Jericho could have been a diversion that allowed Israelite "special forces" to infiltrate the city by way of the harlot Rahab's house adjacent to the wall, we may find the *Bible* a reasonable source for how that battle was won.[1] If so, we may consider again the case of Gideon, who with 300 picked men defeated a much larger Midianite army through stealth and surprise, attacking in such a way that the Midianite force was thrown into great confusion.

Throughout his writings, Sun Tzu emphasized the preeminence of command over other factors. "Therefore the skilful [sic] commander takes up a position in which he cannot be defeated and misses no opportunity to master his enemy."[2] "To be certain to take what you attack is to attack a place the enemy does not protect. To be certain to hold what you defend is to defend a place the enemy does not attack."[3] Had Sun Tzu's advice been available to General Custer, for instance (it is at least possible, since *The Art of War* first appeared in the West a century before Custer's Last Stand, having been translated into French by a Jesuit priest in 1772), and had Custer heeded that advice, he might have approached the Little Big Horn River differently—or not at all without waiting for reinforcements.[4] There is little evidence to suggest that Custer knew of Sun Tzu's admonitions, however, and none that he heeded them.

We might consider any of dozens of clashes of huge, similarly armed forces. Keegan asserts that the defeat at Agincourt was at least partially due to a "lack of effective overall command in the French army."[5] Can we find other examples in which one commander made better use of his forces? For instance, was Wellington equal to Napoleon? Or did he, in the key instance of Waterloo, perceive and react more astutely than the Emperor? Considering that the field was shrouded in literal fog that day, and that Napoleon "had chosen to take no part in [the battle's] tactics" but "was watching from the height on the far side of the valley" while Wellington rode back and forth across the breadth of the line as the action moved from point to point, the Duke was certainly in a much better position to evaluate the situation and respond to it.[6]

What of other inputs to command? For instance, how useful is intelligence if it cannot be used? Can we find an example where a leader knew where and/or when the enemy would attack but was unable to use that knowledge effectively?

This matter of available intelligence became the subject of many conspiracy theories after the 11 September attack; and, much like the Pearl Harbor attack, also drew the interest of Congress. Without going into detail about the specific incident, we can postulate several general "what-if" scenarios regarding foreknowledge of enemy attacks (leaders did or did not know of enemy intentions and plans in advance), action to prevent enemy attacks (leaders did or did not take action), and the ultimate success of any attack. The results of the interactions between these factors may produce quite different popular reactions and conceptions, from the citizens believing the attack was only miraculously thwarted to their leaders failing to act prudently. Eight different interactions are shown in Table Five.

Table 5. Interactions: Foreknowledge of and Action Against Enemy Attack. The "reactions" listed will not represent the entire populace, but they may be expected to surface among the citizens.

Leader Knew	Leader Acted	Attack Succeeded	Expected Popular Reaction
No	No	No	Saved by a miracle
No	No	Yes	Should have known and acted
No	Yes	No	Fortunate; perhaps even lucky
No	Yes	Yes	Should have known or done more
Yes	No	No	Should have acted . . . negligent?
Yes	No	Yes	Should have acted . . . traitorous?
Yes	Yes	No	What leaders are expected to do
Yes	Yes	Yes	Should have acted differently

The reaction of a populace to an enemy's success may be expected to change depending on how much their leadership knew and whether the citizens deem any actions taken to have been adequate. Unfortunately, some of us seem to attribute some degree of omniscience to our government when it comes to what our enemies may be planning—or perhaps we have unrealistic expectations of our government's ability to foresee and forestall the actions of our adversaries. Our government's limitations in this regard are all too evident, however; for instance, during the months leading up to World War Two, the US government failed to make intelligence reports that the Japanese were planning an attack available to the right people, and thereby missed several opportunities to head off disaster.

The role of intelligence in early submarine warfare also seems to qualify it for Case 3. German U-boats, for instance, enjoyed a great deal of autonomy of command and control. Their weaponry was limited, and their command functions were hampered by the difficulty of finding and fixing on enemy shipping, but they enjoyed a considerable tactical edge. In contrast, their adversaries were at nearly a complete loss, and the U-boats enjoyed significant success.[7] Once the convoy fleets changed their tactics to cope with the submarine threat, however, the situation changed dramatically (see Case 4).

World War Two offers us other examples of intelligence success stories that apply to this case. From the capture of German Enigma code machines to the Navajo Code Talkers, the ability to understand the enemies' intentions while keeping our own plans secret was a tremendous advantage in command.

From late in the last century, this case may also apply to the coalition victory over Iraq in Operation Desert Storm. Certainly the US and its many partners combined effective command with military strength; Iraq, on the other hand, had military strength in abundance but its command functions were limited by ineffective information-gathering and, after the first waves of air strikes, loss of many C^2 nodes.[8] This case does not hold, however, for the two preeminent conflicts of more recent years: Operations Enduring Freedom (Afghanistan) and Iraqi Freedom will be considered in Case 5.

Looking to a future combat arena, we might examine US space capability versus a non-space-capable enemy. In the current environment, space is yet to be weaponized—it may, however, be said to have been militarized—and therefore the momentum element is lacking for all participants. Our plethora of space assets, however, gives us a clear illustration of one nation holding considerable advantage in command-related resources (e.g., satellite communications and data-gathering) over other nations not so equipped. For the time being, this is only an advantage in command, because it is unrealistic to believe that space will remain free of all weaponry forever.[9] But this advantage in command is not likely to last. The balance may be shifting even now; e.g., as commercial remote-sensing data become more widely available, commercial satellite communications less expensive, and other powers pursue their own capabilities (such as the Europeans' Galileo satellite navigation system). Unless the space- and ground-based infrastructure is maintained and improved, the US advantage in this area will erode over time.

34

CASE 4—SPLIT ADVANTAGE

It is difficult enough to examine cases where forces are separated along a single axis (either command or momentum). What, then, of forces who hold the advantage in one element but not the other? Can we find examples where one side was "materially" stronger, but the other had stronger command? Do these examples all teach us the same lessons?

Consider Alexander's victory over Darius at Gaugamela. Darius might be said to have had a "decisive" advantage in momentum, outnumbering Alexander's troops by a wide margin. Alexander apparently held the advantage in command, and it turned out to be decisive as he rendered the Persian chariots ineffective and routed Darius's superior forces. The same may be said for Frederick the Great's victory at Hohenfriedberg in 1745, where Frederick had quite good intelligence despite being outnumbered and outflanked: as described by John Keegan, "It was [Frederick's] own quick thinking in the heat of action and the fierce reaction of his soldiers which turned the tide of battle."[1]

It may be that Gideon's victory over the Midianites belongs in *this* case rather than the previous one. And while the relative strengths of the armies are not clear, the Chinese general Sun Pin—a descendent, so it is said, of Sun Tzu—used a clever ruse to lure his enemy into a trap which eventually led to conquering an entire army.[2] Sun Pin ordered the army to build 100,000 campfires the first night, 50,000 the second, and

30,000 the third, to make it appear that most of the army had deserted; as a result, his overconfident enemy left behind his heavy infantry and followed with only lightly armed troops which proved unable to secure a victory.

When Robert the Bruce led a mob of Scottish revolutionaries against the English at Bannockburn, the traditional emphasis on "mass" would have predicted doom for the separatists. The Scots were outnumbered, but through sheer determination—a quality that armies often develop in response to powerful, charismatic leaders—they won their freedom. As Keegan writes, "However good the intelligence available before an encounter appear to be, the outcome, given equality of force, will still be decided by the fight; and, in a fight, determination, again given equality of force, will be the paramount factor."[3] Put another way, when the weaponry is equal the battle is often decided by sheer savagery;[4] how much more so, then, when the forces are unequal.

This case of mismatched advantages also seems to apply to the already-mentioned Boer War, when the Boers, beaten back from Mafeking and elsewhere, turned to the stealthy commando units to prosecute the war. The British forces, with superior strength, found their command and control hampered by lack of intelligence on the activities of the commando; in contrast, the materially weak commando forces were able to gather sufficient information on British strengths to stage several successful, albeit uncoordinated, raids. While this shows that superior command can win some victories against superior momentum, it is important to note that the commando, although successful tactically, did not effect a strategic victory for the Boers.

The quintessential example of this case may be the battle between the Germans (with the command advantage) and the Russians (with a strong momentum advantage) at Tannenberg. As explained by Liddell-Hart, the German troops,

> inferior in strength to the Russians, could not have been decisive; but Ludendorff . . . took the calculated risk of withdrawing the rest of the German troops, except the cavalry screen, from that front and rushing them back against Samsonov's right wing. This daring move was aided by the absence of communication between the two Russian commanders and the ease with which the Germans deciphered the Russian wireless orders. Under converging blows, Samsonov's flanks

> were crushed, his centre surrounded, and his army practically destroyed.[5]

Note that Liddell-Hart states that the German troops "could not" have been decisive. Obviously, they were. That original (and erroneous) judgment was based on the fact that the Russians held the numerical advantage, a strictly Jominian interpretation that ultimately was incorrect.

As mentioned in the previous chapter, if Case 3 described the early days of submarine warfare, Case 4 more describes the near parity that developed shortly thereafter when new tactics were applied. Convoys grouped supply ships together with sufficient escorts to thwart many attack attempts or render attacks less effective. The large convoys with escort vessels represented a change in momentum, giving Allied shipping an advantage over the German U-boats, who retained only a slight edge in command. If we look a bit deeper, however, this example emphasizes the difficulty of applying strict categories to dynamic and complex situations because the tactic of convoying may be considered a change in command that resulted in a change of momentum.

Keeping with the nautical theme for the moment, two of the pivotal Pacific battles of World War II also seem to illustrate this case. At Pearl Harbor, the Japanese task force did not match the material strength of the US force, but their application of command to achieve surprise overcame the relative lack of momentum (although it may be that the Japanese momentum—their combat power in motion—was greater than the US fleet's combat power at rest). At Midway, the Japanese strike force had excellent strength, but their command ability was compromised by an intelligence breach: decryption of Japanese codes, and subterfuge by US signals operators, revealed the Japanese fleet's final destination. On the US side, fairly good strength (though inferior to that of the Japanese) combined with superior intel to balance the fight. Midway is another difficult example to pin down, though: The US did not muster truly coordinated action throughout the battle, and ultimately relied on "luck and shrewd guesswork" to pave the way to victory, so the US may not really have had a true command advantage after all.[6] Keegan again: "The events of 4 June 1942 at Midway [show] that, even when intelligence seems to provide the explanation of a victory, closer examination of the facts may reveal that some other factor, in that case chance, lies at the root of the matter."[7]

This case of split advantage may also apply to the early days of the 1948-49 Arab-Israeli War, during which the Israeli advantage in command coupled with fierce determination overcame the initial superiority of Arab numbers and firepower long enough for the Israeli forces to achieve parity and eventual superiority. The 1967 Arab-Israeli War followed a similar pattern of command advantage, this time through superior intelligence and application of available weapons through the element of surprise.

Similar to the Boer War discussed above, this case may apply to wars of insurgency or rebellion against strong forces. The archetype may be the activities of the Viet Cong against South Vietnamese and US forces. As with the British in South Africa, and the French in Vietnam before, the US often lacked basic intelligence on the adversary's abilities, plans, and activities, impacting overall command functions. The Viet Cong, on the other hand, combined a distributed but effective C^2 system with good information about US strengths and vulnerabilities to make up for their overall military weakness. By the time the focus of the war shifted from an insurgency to a more conventional fight, however (e.g., in the Tet offensive), this case no longer applied.

Renowned historian Martin Van Creveld asks, "How can one explain the victories won by guerrillas who, initially at least, possessed no military experience or organization, hardly any weapons, very limited economic resources, and (in some cases) not even shoes?" Van Creveld's question paints a picture of combatants weak in both Momentum and Command, but his answer indicates that what initially appears to be a disadvantage in both factors may not be so:

> The more powerful and more modern the technology at the disposal of an army (and the more modern, therefore, the organization, training, doctrine, and command system built around that technology), the less useful it was in combating an enemy who did not represent a territorial state, did not have permanent bases or lines of communication, did not possess modern weapons, and, most important of all, could not be distinguished from the surrounding population.[8]

In other words, the Command structures of the modern army are optimized for a different kind of war—which has become known as Fourth Generation War, or 4GW—and the guerrillas' adaptability to the exigencies of low-intensity warfare may give them an advantage in useful Command.

Revolutionaries around the world and throughout history—British colonists in the Americas, Democratic populists in France, Communists in Russia, Maoists in China, Afghani freedom fighters—have enjoyed this split-advantage situation:

> Guerrilla warfare was the most dreaded form of operation for a regular army for both military and political reasons. As noted, from a military standpoint, regular armies, even those with substantial colonial experience, were poorly equipped to deal with it. An elusive enemy could control the strategic pace of the war, withdraw deep into the country, and nullify the technological and firepower advantage which should naturally be enjoyed by the invaders.[9]

The lack of material strength does not seem to hamper revolutionaries as much as the lack of effective intelligence hampers those they rebel against. It is a pattern very likely to repeat, and futurists have even suggested a drawn-out guerrilla campaign as the most likely means to contend with any invasion of Earth by a technologically-superior alien race.[10]

Nations do not always learn these lessons, however. Since Versailles, many conflicts have ended with contingents of troops in place to enforce the peace. Often these were not conquerors in the purest sense. They were not there to hold new territory that would now be considered part of the conqueror's nation, but as police to keep the still-sovereign locals in line, which seems a strange way to end a war. Such an end is certainly less than what might ever be considered "total victory" in the sense discussed with respect to the victory/defeat space presented in Chapter 24. It seems like a failure of grand strategy, a failure to plan for the peace that will follow the conflict.[11] Yet this pattern continues. The "victory" of NATO forces in Kosovo resulted in another quasi-occupation; so too the quasi-occupations of Afghanistan and Iraq. Even though the bulk of US forces were withdrawn from Iraq, can we say we truly "won" that conflict? And at what point, if ever, will we be able to say we "won" in Afghanistan?

It seems final victory must wait until we no longer have to stay in a place to enforce a peace that is only tentative, or perhaps until we are *invited* to stay to help protect a newly democratic allied nation from its neighbors more than from agitators within its own borders (as we have done in the Korean peninsula for decades). We may hope to maintain the national will to see current and conflicts through to such a victory.

As we have seen, there are cases where good command and control can make up for poor strength. Often this is a short-lived advantage, or an incremental one that must be sustained for a long time to effect victory (as in revolutions). But it is an advantage that can be exploited, leading us to question again under what circumstances the long-standing emphasis on brute strength—in the conventional wisdom of mass and concentration—prevails in the real world. It seems, in the end, to win out primarily in wars of attrition or annihilation, which may favor the US since such operations may be considered our national modus operandi.[12]

35

CASE 5—DOUBLE ADVANTAGE

In this case, where one side holds the advantage in both command and momentum, we should find the clearest examples of overwhelming victory. But do we?

The supreme Twentieth Century example of the double advantage would seem to be the German invasion of Poland in September 1939. German forces so completely overwhelmed the Poles that the invasion was completed and the entire country captured in less than a month (including the ten-day siege of Warsaw). This same case may also include the subsequent German invasion of France, even though the French army was better equipped than commonly imagined.[1] French commanders were not sufficient to the task of repelling the Germans, and French intelligence did not foresee the German advance through the Ardennes. Once the advance began, French forces were tied too closely to the Maginot Line—in spirit if not in absolute fact.

If World War Two in Europe began with a double advantage on the side of the Axis, the war in Europe and the Pacific ended with a double advantage firmly in the Allies' grasp. The combination of Command—planning, secrecy, and deception—and Momentum—firepower, mobility, and the critical elements of human determination and endurance—won both theaters by grinding our enemies down bit by bit.

Do we find any contrary examples that fit this case? Any examples in which the side that *lost* had both superior command and superior momentum? Where a thoroughly weak army vanquished a strong, resourceful, and well-led force?

We have discussed previously the efforts of the Viet Cong, and some analysts might fit them into this case. We have also discussed the asymmetrical resistance afforded by native Indians to British rule, and some might argue their eventual victory belongs here. Again, while the concepts are easy enough to define and grasp, they are broad enough to be open to interpretation and so we may not arrive at a final answer.

Sticking with the conceptual argument, however, it is difficult to conceive of a case where the victor held *no* advantage whatsoever, because every successful attack or defense appears to succeed precisely because the victor makes use of an advantage—even if that advantage, by deception or ignorance, appears like a weakness. As Sun Tzu wrote, "Apparent confusion is a product of good order; apparent cowardice, of courage; apparent weakness, of strength."[2] A completely disadvantaged victor, then, is almost a contradiction in terms.

36

ELEMENTAL QUESTIONS

Do the five cases we examined in the last five chapters explain all aspects of the historical victories mentioned? No. Can the cases be relied on to predict victory in every postulated future conflict? No.

Why not? As Sun Tzu wrote, ". . . it is said that one may know how to win, but cannot necessarily do so."[1] Unfortunately, and this is a basic tenet of Theory of Knowledge, the outcomes of battles are not perfectly predictable. In the field of military theory, we remember that the uncertainties of battle take the form of fog and friction, those perennial and unconquerable enemies that may combine to make all this moot, or at least to increase the difficulties of applying this or any other concept in a coherent manner.

In the quest we are now completing, we derived two elements that we may consider to be *essential* to struggle and conflict: combat means, which we termed "momentum," and the method of directing those means, which we called "command." In every encounter we studied or postulated, some Momentum and Command were applied, although at times they took different and unfamiliar forms.

Can we say with confidence (never with certainty) that, if we have the strongest Momentum and the most robust and capable Command, we will always be able to fight and win? Perhaps, but the situation is never static. We must ensure that we select and apply the means of war —that we *use* our momentum—in ways governed by clear understanding of the strategic and tactical situation. Even then we must adapt to the dynamics of the conflict—the battle rhythm—to maintain any advantage we might have had at the start.

Understanding that war is an act of will, the element of command is of prime importance. What the commander sees, whether through binoculars or the electronic eyes of a satellite—and how he interprets what he sees. What the commander hears, whether over a telephone or radio or an Advanced Extremely High Frequency satellite—and how he interprets what he hears. What the commander *knows*, in the fullness of that meaning, and the strength of the commander's determination to conquer the enemy. Command ability, the elusive quality sought after by every military practitioner and every military theorist, is still the Holy Grail of war and warriors.

Do we have better situational awareness than our adversaries? Are our plans kept secret? Have we telegraphed our real moves, or are our deceptions working?[2] We would do well to heed Wang Xi's advice, in commenting on Sun Tzu: "Consolidate your keenest energy, save up your extra strength, keep your form concealed and your plans secret, being unfathomable to enemies, waiting for a vulnerable gap to advance upon."[3]

Do we have better people—more highly skilled, better trained—than our enemies? Do our people have the will to fight and win? If so, do our commanders have the skill and will to employ them effectively? If not, how (and how quickly) can we bring them up to fighting form?

Do we have better supplies in place, and better supply lines? Do we have better weapons—type or amount—and transport? At the outset of the second World War, the US in particular did not; by the end of that war, not all of our arms were materially better (in fact, some were inferior), but we had far more of them (an example of "mass") and could move them at will (producing the element of Momentum).

Based on our understanding of the enemy, can we predict what they are likely to do? If so, we may be able to discern what actions are likely to lead or coerce the enemy into behaving to our advantage. If not, and we fool ourselves into thinking we know how they will respond, we may find ourselves in the enemy's trap—like the French in 1940. Will the enemy move or react quickly? Will they split their forces, or concentrate? Will they retreat, feint, or engage—and if we attack, feint, engage, or retreat, what will they do in response?

And always: How confident are we in our predictions?

As we formulate answers to these and a thousand other questions,[4] too many different ways of studying and understanding wars and battles present too many choices to the student of war. Our traditions may lead us to apply the Principles of War, but if we do so, which ones? The nine traditional Principles, or some expanded or truncated list? Do we apply them selectively, or collectively? What other factors do we consider besides the canonical principles? Do we apply the tenets of airpower? The "Five Rings"? The physical, mental, and moral aspects? Insofar as this text is concerned, our attempts to answer these questions about ourselves, our enemies and our situations often confuse as much as they clarify, especially as we try to absorb so many different theories—e.g., to take into account the level of war involved in the conflict, the media in use (or useful), and the multitudinous characteristics of the enemy and our own forces. Instead of producing the keen perception and quickness of understanding of *coup d'oeil*, they may leave us mentally exhausted, and may even produce an unwelcome quality denoted by another French term: *ennui*.

In the end, despite so much that we might know and so much more we will never grasp, if we are to accomplish anything we must *act*—and since our knowledge is never complete we will act at times without knowing fully why, or fully what we expect.

Our knowledge must always remain limited, so it is clear that we have not built a perfect model in these elements of Momentum and Command. We presented them to complement—but not to replace—the various theories we discussed in this text, and showed that these basic elements are necessary to conflict with respect to the levels of war, media of war, and other theories. That at least makes them more universal than some other descriptors, so it may be a somewhat *useful* model. Knowing that these two elements must be a part of any conflict at any level, and that a continual advantage in both may contribute to victory, we may be able to focus on developing and exploiting any advantage we hold. Based on the advantages we perceive, we can apply our energy and creativity to pressing these advantages home and compensating for any disadvantages. The end result should be, if we have analyzed the situation correctly, a greater probability of victory. Sun Tzu understood this when he wrote, "Know the enemy and know yourself. . . ."[5]

With respect to knowing ourselves, several times throughout this text we discussed the "will" to fight. With respect to the commander's authority, to the reasons for conflict, and to the troops' determination to fight and continue fighting, will enhances both Command and Momentum. Troops who lack the will to fight will not build warfighting momentum; commanders who lack the will to fight will be unable to make effective command decisions. This aspect of morale works in strange ways, also: strong commanders can affect troop morale, troop morale can affect commanders' morale, and (especially in democratic nations) the will of the populace and the will of the civilian leadership can have tremendous effects. For example, consider the March 2004 Spanish elections that were tipped toward the challenging party after Al Qaeda's bombing of a commuter train in Madrid. The resultant withdrawal of Spain from the coalition fighting in Iraq confirmed for the bellicosi the effectiveness of their tactic (attacking civilians going about their daily business) and their broad target (a nation holding elections). On the other hand, it may be that more recent election results following waves of predominantly Muslim immigration into Western nations will make it more difficult for the bellicosi to infiltrate and operate in those societies.

With respect to maintaining the will to fight and win, clear objectives and just causes—as in World War Two—enable troops, commanders, and whole populations to make terrific sacrifices in order to obtain victory. The will to fight was both a force multiplier that increased the fighting momentum and a catalyst that crystallized the commanders' determination and resolve. In conflicts with fuzzier objectives, however—and especially conflicts that have ambiguous goals —both command functions and available momentum suffer. Therefore, if we are to maintain the collective will to fight and win the larger ongoing Terror War, it is imperative that its objectives remain clear and just. With that in mind, perhaps the biggest weakness of this Command and Momentum construct is that it does not explicitly account for such considerations as the reasons for the war itself, whether taken on the strategic or grand strategic scale.

In the end, then, we are not close to a unified theory of warfighting, though we may have learned something of how knowledge and theory interrelate and how theories of war may be tested against experience. We are faced with two possibilities, each at extreme distance from the other. First is the possibility that nothing matters: nothing is a reliable predictor of victory or success, so we serve no useful purpose in pursuing these lines of reasoning (or flights of fancy). Second is the possibility that everything matters, in which case the only way to reliably predict victory or success is to learn and apply everything. The first instance requires gifted leaders; the second, genius leaders. As with most paired extremes the truth lies somewhere in the middle, which is good news for those of us whose gifts or genius are mediocre at best.

37

THE SEARCH FOR NATURAL LAWS OF WAR

This entire book has been based on the premise that Theory of Knowledge, as a way of thinking about the world and the way the world works, could be applied to war and warfighting—and therefore teach us something about war. But what does it teach?

According to the Theory of Knowledge we examined earlier, all knowledge is prediction, which we can illustrate with the proposition, "X will happen." Usually we understand this to be based on inputs and prevailing circumstances; in the same terms, "if A, then X will happen." It would be more complete to recognize that the determining factor "A" involves a *huge* number of variables and therefore state the proposition using the sum of those inputs: "if $[A_1 + A_2 + \ldots A_n]$, then X," or "if ΣA, then X." In the realm of warfare, what our scouts and spies tell us of the enemy, the terrain, etc., are *data*; we interpret those data to produce *information* by which we understand the enemy's arrangement and potential, the best avenues for approach and escape, and so forth; but only when we predict where the enemy will attack (or whether, or when) or how they will respond to our actions have we produced *knowledge*. In order to repel the enemy or advance successfully against him we need knowledge, not raw data or even information—though these latter components form some of the building blocks of the knowledge we need.

In the military context, we are concerned with predicting victory, which we may illustrate by another proposition in the form, "if A, then V." In this case the situation becomes much more dynamic, in that one of the large number of variables in A must be an inherent prediction of what one's adversary is planning or likely to do. Therefore, the representative statement for this proposition becomes more complex: "if Σ[A + (if A, then E)], then V." The situation quickly becomes so complex that only an incremental or iterative process can resolve these nested predictions, where every tick of the clock or flicker of the digital display brings new information on current status, enemy formations and movements, weather, etc.

Every sound and useful military theory, then, if it conforms to Theory of Knowledge, will predict future outcomes with some measure of confidence (if A, then V) or explain with clarity events of the past (V, because of A). Theory of Knowledge also teaches us that nothing is predictable with certainty—all our predictions and all our knowledge have the possibility of error. That is part of the reason why the cases examined in Chapters 31 through 35 could not resolve every example with perfect clarity, and part of why it was so difficult to find examples that fit neatly into the cases in the first place.

Theory of Knowledge teaches us that we run significant risks when we try to understand war only from the perspective of history—that is, from the perspective of either direct or vicarious experience. Theory of Knowledge shows that we must interpret experience in the light of theory. One over-used and over-simplified example that we alluded to earlier may suffice to illustrate this, if nothing else because of its familiarity. The French understood the *history* of trench warfare and German aggression, and built the Maginot Line, though Liddell-Hart argued that their "incomprehension of the new idea of warfare" was more profound: "What proved fatal to the French was not, as is commonly imagined, their defensive attitude or 'Maginot Line complex', but the more offensive side of their plan. By pushing into Belgium with their left shoulder forward they played into the hands of their enemy, and wedged themselves in a trap. . . ."[1] The Germans, on the other hand, understood and refined the *theories* of combined airpower and fast armored warfare, bypassed the Maginot Line using an updated version of the Schlieffen plan, and drew the French into the trap that caught them.

Our theories and concepts present us with a framework within which to place experience, together providing a basis for decision-making. Those decisions still have the possibility of error: they may be right or wrong based on what is known at the time. Furthermore, those decisions are uncertain in their outcome: they may turn out to be good or bad, and not based entirely on controllable factors. And those decisions —*our* decisions—become inputs to still further explanations, and theories, and decisions.

Because all knowledge contains the possibility of error, which in warfare is further complicated by fog and friction, the aim becomes increasing the probability of victory rather than assuring the certainty of victory. With that in mind, the ability to predict, if not actually to control or drive, the adversary's actions, becomes an important input into the overall equation. We may be able to predict with some confidence the enemy's likely reactions to what we present him; or, we may be able to entice the enemy in a certain direction or freeze him "on the horns of a dilemma" as in Liddell-Hart's indirect approach.[2] Whatever we try, we will never be certain of victory—though we may grow to be somewhat more confident of it.

Our simplified theory of war postulates that every war, every battle, every skirmish, every fight, involves the application of some combat means through the (hopefully) conscious decision of the combatants. Can this or any other martial theory eventually be accepted as a "law" of war?

Others have made such claims. Echoing Jomini's faith in his over-arching principle, Mark McNeilly wrote in 2001, "The [six] principles I have culled from [Sun Tzu's] *The Art of War* are much like the laws of physics; they exist whether we know of them or not. A scientist or engineer acting in accordance with the laws of physics, is successful. The same is true of a commander; using these principles of strategy, he too is successful."[3] Note that McNeilly left little room for doubt or error; in contrast, we maintain that both doubt and error must still be in play according to Theory of Knowledge.[4]

How does a theory become accepted as "law"? When a theory proves reliable in predicting or explaining experience in every instance over a long period of time, it usually advances in status to that of a law.[5] The best known "natural" laws are so overwhelmingly consistent they appear never to fail.

The most familiar of these "natural" laws is the law of gravity. Even if the apple story is apocryphal, Sir Isaac Newton initially formulated a theory to explain the phenomenon of falling objects. That theory is now considered a law since it never fails in our Newtonian workaday world. Other familiar examples are Newton's laws of motion and the laws of thermodynamics. In all these cases, if we were to observe a phenomenon that defied the law (e.g., an object repulsed rather than attracted by gravity, or a process in which energy and matter were not conserved), we would have to re-examine the basic theory and either modify or reject it. Since we do not observe any failures in these theories, we may call them "laws" and use them as bases upon which to predict future outcomes. Based on such stable, reliable laws, our predictions should have a much higher probability of being accurate.

Some theories, if not most, probably never will be known as laws, either because they undergo continuous refinement as evidence is collected or they are supplanted by later theories that more accurately reflect experience. The theory of evolution is an example of the former, because biology and paleontology are continually adding new insight into the processes and mechanisms by which life develops and changes.[6] As an example of the latter, new developments in quantum theory may preclude the Theory of Relativity from becoming a law, or may modify it significantly before it reaches that status.

However, even a theory that acquires the status of a natural law may be further refined. For instance, our understanding of gravity and other aspects of Newtonian physics was influenced by Einstein's theories—especially the General theory. And though Newtonian physics breaks down at the sub-atomic level, we do not live our lives there: the everyday "laws" of motion still govern the world of our existence. This flexibility allows us to refine theories but hold on to those general cases which we need to survive; as C.I. Lewis wrote,

> A 'law' having a high degree of probability may have to be abandoned in the light of future experience as not universal, and may yet remain a 'rule of thumb' or statistical generalization which is true in the great proportion of cases and hence gives rise to a still valid probability in the particular case.[7]

Thus even the Principles of War can remain useful even though they cannot form a complete theory of war.

Whether these characteristics of natural laws ever allow us to compose over-arching laws of warfare remains to be seen. The construct of Momentum and Command we developed herein certainly does not qualify; but it is doubtful that any military theory ever will.

38

THE RISK OF WAR, REVISITED

The first time we considered the risk of war, in Chapter 3, we presented "men will fight" as a prediction, and further predicted that "men will fight wars." Accepting that proposition allows us to posit two questions: How will men fight wars? and why?

Our main concern in this volume has been "how." Not the details of tactics and weapons, but from a different vantage point: Men will fight wars smartly or stupidly. Unfortunately for us, it is too simplistic to say that fighting wars smartly will always lead to victory. Warfighting is a messy, muddy affair, in which smart fighting may still result in loss. Victory may sometimes reward dumb luck, but we wish not to rely on dumb luck.

We are partly concerned with "why," in that the reason behind the war may influence how it is fought: covertly or overtly, alone or as part of a coalition, etc., but always over things we *value* (as stipulated in Chapter 8).

We recognize that nonviolent conflict is not only possible but prevalent in other realms. Diplomacy often seems to be the art of nonviolent conflict. Do the elements of war obtain in these other realms? Diplomatic Command—the control over and conduct of relations of ambassadors, attaches, and envoys—seems a workable concept. Diplomatic Momentum, on the other hand, is a fuzzier concept; it would possibly consist of the threat or application of demarches or other censure actions, but applying violence-referent terms to what should be a nonviolent activity may stretch the idea too thin. In the economic realm, Command is more nebulous in a capitalist economy that tends toward laissez-faire; in the US, for instance, the Federal Reserve exercises some control, but usually at a distance. Economic Momentum is easier to describe, as it may consist of everything from loans and foreign aid to sanctions—the latter of which may be taken as a form of economic violence enacted to convince or coerce the sanctioned to accede to our will. In the end, applying these constructs to other realms may best be done under the aegis of Grand Strategy (recall the discussion in Chapter 16).

We may wonder if the elements of Command and Momentum can aid decision-making when lives and honor are at stake; i.e., whether they can help us decide whether to attack, and when. If we consider them in light of our enemies' capabilities, perhaps so; if we can take action that will adjust the situation in our favor, perhaps more so. Of the two elements, though, Command must be preeminent in short-term or immediate decisions, since developing the combat power that forms the basis of Momentum must take a finite and often lengthy amount of time. As to the decisions to be made, it is overly simplistic to say that one should attack when one has the advantage in both elements. While that may increase the probability of victory, it does not address the temporal aspect of Theory of Knowledge; i.e., the aspect of prediction and the consideration that all knowledge and predictions are probable rather than certain. Here again is why the Command element is key in decision-making: because only Command can determine if the expected payoff of attacking is worth the cost and the risk.

These decisions and considerations rise to the arena of strategy and grand strategy. Unfortunately, grand strategy, as a concept and a practice, is not widely accepted or understood in the US.[1] It is almost as if the US does not believe nations should have wide-ranging and clearly stated objectives; however, our reluctance to tackle grand strategy may have simpler explanations such as the general reluctance of democratic societies to make war. Perhaps our natural and historical isolation leads us to respect (or, at times, to tolerate) other nations and not wish to impose our will on them. Perhaps simple selfishness is partly to blame: Often our domestic concerns and economic interests overshadow any international aims we might have, and we effectively ignore other nations until problems abroad require our attention, if not our assistance or intervention. It is more likely that our pluralism keeps us from developing a set of coherent national goals; in addition, our preference for peace may keep us from believing (or maybe admitting) that some national goals are best met by waging war.

But, as Sun Tzu wrote, "War is a matter of vital importance to the State. . . . It is mandatory that it be thoroughly studied."[2] While US defense spending remains robust even at a low percentage of the Gross Domestic Product, the US government and US culture itself has slowly but steadily de-militarized—evidenced by the relatively low numbers of elected officials and other public servants who have served in the military.[3] This might at first indicate that we no longer recognize Sun Tzu's statement as true. More likely, though, it is a function of our success at infusing warfare with technology such that fewer military personnel may wield greater combat power, whereby we have fewer veterans to serve in public office or civil service. The end result is the same, however: The study of war, rather than being important for entry into public service, is now the province of a professional warrior class whose civilian masters, because they have not studied and do not understand war, may actually be too quick to wield the sword when other instruments of power are still available.[4] This must be unintentional, for no other reason than lack of experience; as historian Colin Gray wrote, "Politicians may know too little, if not of war *per se*, at least of the probable character of war today, to know what to demand or expect of their generals."[5] The corollary question, which we may not want to ask and answer, is whether we are in danger of growing a military elite as our population grows and our military shrinks.

In the early years of the 21st Century, our campaign against the bellicosi naturally captured our attention and the attention of much of the world; however, we should bear in mind that other battle arenas have not been thoroughly explored or exploited. Widespread electronic, or "information" warfare, potentially can affect the other traditional instruments of power. Electronic attacks could disrupt civil government functions, interrupt power transmission, ruin commercial transactions and records, etc. The "virus," "worm," and denial-of-service attacks on computer systems in recent years, as annoying as they have been, have usually not approached the level of sophistication and devastation that is possible: consider the effects of the Stuxnet virus on Iranian uranium centrifuges and how similar attacks may affect other systems. The use of various electronic media by the bellicosi, including their use of steganography[6] to send encrypted data, indicates that they have learned to use them for their own purposes; if they grow more adept, they may attempt to disrupt the electronic arena as readily as more technologically advanced adversaries. Even though they seem to prefer traditional tactics that produce shocking scenes of mayhem, we may be advised to consider them even more likely to exploit and damage networks if they perceive they can create large disturbances with minimal investments. And the recent reports of Russian opinion-shaping and influence-peddling in the US electorate through targeted social media campaigns revealed yet another avenue of approach for enterprising adversaries.

The expansion of conflict beyond the traditional battlefield into the electronic arena may require us to paraphrase Master Sun: "War, both real and virtual, both physical and informational, is of vital importance to the state." As in conventional war, electronic and information warfare require Momentum and Command, even though the means of combat may not fit common descriptions and the mechanisms of command may be hard to implement. But "means" and "methods" may translate better into computer terminology than "tactics" or "logistics," so the theory discussed in this treatise could prove more useful to electronic warriors than other common theories.

Regardless of the mode of warfare, whether it is conducted in the real world or some virtual world, war has consequences for real people in this world. If we are to mitigate the risk of war, and reduce either its probability or its severity, then we must study it diligently and seek new ways to understand it. That, after all, is why we advance possible new ways of thinking about this ancient subject.

It seems safe to predict that the construct presented in this volume will not enjoy wide acceptance, as it is neither as simple nor as complete as we may have wished; it is certain, as noted in the last chapter, that it will never reach the status of law. In addition to the possible errors that we enumerated in Chapter 21, this entire methodology may be backward. Proposing two "elements" of war may be as useful to conducting a real military campaign as the classical physical elements—earth, air, fire, water—are to modern chemical engineering.

One price of modernity, and of a modern outlook on the world, seems to be acceptance of higher and higher levels of complexity. As mankind's understanding of the world grew, we discovered new elements and learned more about the building blocks of everyday and exotic compounds. Water became dihydrogen monoxide, salt sodium chloride, etc. From that standpoint, we could just as well have looked at this subject of martial elements by reduction:

Water molecule → Hydrogen & Oxygen →
Protons, Neutrons, Electrons → ???

Combat means → weapons, troops →
M16s, M4s, riflemen → ???

Command → situational awareness, experience →
reconnaissance, training → ???

It might have been more complete, then, to build a Periodic Table of War, populated by everything from weaponry to logistics to communications to training—and from which all the possible "molecules" of conflict may be built. Unfortunately, that model would suffer because even the Periodic Table has been subject to revision, both by the addition of elements and by new ways of depicting their arrangements. Such a construct would also be much more complex than the model presented here, and we have seen the disadvantages of complexity in Chapters 22 through 25. Complexity does not lend itself to expedient decision-making in a compressed timeframe; plus, it is important to remember that we have been discussing general ideas applicable to all conflict, not specifics that may work in one instance and not another.

Recall the question from the first paragraph of Chapter 1: The skeptic's privilege is to ask, "How do we *know*?" At least part of this quest has been to determine whether what we think we know about war is true. In some cases, we showed widely accepted ideas to be questionable, and knowing the limitations of those theories should be important to those who have to apply them. In others, we combined widely accepted ideas in new ways, some of which may give military practitioners new insights or ideas.

As we have pointed out repeatedly, Theory of Knowledge teaches that to know anything is to make predictions about the future. To know any *thing* is to make predictions about the future of that thing—not to describe, but to interpret, to prophesy. Recall that we know the object not as a copper disk but as a penny; as a military analogue, it is not a "tapered brass cylinder," but a 5.56-millimeter round. Even to call someone a "soldier," rather than a "person," implies a certain possible or potential future that is true of one but not the other. This mode of thinking, as C.I. Lewis showed and we studied in Chapters 4 and 5, is as natural to intelligent thought as breathing is to life.

The answer to the skeptic's question about war is—the *truth* is—in many cases, we do not know. As C.I. Lewis put it, "Knowledge has two opposites, ignorance and error. The relativity of perception may mean ignorance or it may not."[7] Whether ignorance or error, our conventional wisdom about war—our "common sense" about it—is not as robust as we would like. Many of our theories, where they can be tested against experience, fail: only those theories that pass the test, that jibe with reality, can we accept as adequate predictors. In this volume we tried to make sense of the prevailing theories, and proposed some basic ideas that may hold true across a wide range of experience.

The various mental constructs presented here—the combinations of levels and media, the interactions between principles, the war tree and war as a system, the simplified elements of battle—are unfortunately unwieldy tools, too cumbersome to apply in battle and possibly too cumbersome even for the real world. We should not be too surprised at their shortcomings, however, given the failure of so many other theories to live up to their billing. As Clausewitz wrote, "it would be wishful thinking to imagine that any theory could cover every abstract truth. . ."[8]

Uncertainties in war, the products of fog and friction, are as inevitable as the physical uncertainties of quantum mechanics and the Heisenberg principle. Again, Clausewitz observed, "From the very start there is an interplay of possibilities, probabilities, good luck and bad that weaves its way throughout the length and breadth of the tapestry."[9] In addition, new weapons and further development of the fourth medium of war—space—will necessitate more updates to the theories by which we plan to prosecute war. So, even more in dynamic human endeavors such as warfare than in mathematically rigorous fields such as physics, a unified theory may be an abstraction we are incapable of reaching. We may come closer and closer to it over time without describing it fully.[10] In short, we do not know what we think we know about war.

Sun Tzu warned that it is impossible to discuss specific applications before a particular battle: "When I have won a victory I do not repeat my tactics but respond to circumstances in an infinite variety of ways."[11] Mei Yao-ch'en, another of the many commentators on *The Art of War*, wrote, "When confronted by the enemy respond to changing circumstances and devise expedients. How can these be discussed beforehand?"[12]

Clausewitz warned that "the art of war . . . cannot attain the absolute, or certainty; it must always leave a margin for uncertainty, in the greatest things as much as in the smallest."[13] However, simply because our knowledge is probable only, and our theories always have the possibility of failing to adequately predict the future or explain the past, will not stop us from theorizing. Constructing and testing theories is how we make sense of the world, let alone how we make sense of war. Without developing theories and testing predictions, we have no hope of developing even a trace of *coup d'oeil,* and thereby little hope of evaluating and responding to the circumstances and exigencies of war.

War is, at its heart, a profoundly human enterprise—"the most complex of human endeavors," according to Colonel John Warden.[14] It may not be amenable to precise descriptions as are mathematics or physics, but it still requires a theoretical basis, even when "Circumstances frequently compel political-military leaders to depart from sound theory and principle. Such departures should be made knowingly *and with an understanding of probable consequences.* They should not be made through ignorance or inadvertence."[15] Or, as stated by Jomini, "In this important crisis of battles, theory becomes an uncertain guide; for it is then unequal to the emergency, and can never compare in value with a natural talent for war, nor be a sufficient substitute for that intuitive *coup-d'oeil* imparted by experience in battles to a general of tried bravery and coolness."[16]

In that vein, this is only a work in progress and never will be complete. Even if this elementary construct of Momentum and Command is accurate, Darius would still lose at Gaugamela, and the Spartans would still fall at Thermopylae, because "the real difficulty is that even the best professional military advice frequently will be wrong."[17] The only value of this exercise, and its chief danger, as with the value and danger of all knowledge, will be in the future. Just as martial thought has developed over the years, new methods and theories will rise to supplant present ones; and every theory, even every theory of war, must be tested against all experience yet to come. And even then, as Martin van Creveld notes,

> Theory can aspire to save the strategist from the need to think out everything from the beginning, and provides him with a starting point for thought. Insofar as the theory is sound, such a starting point is certainly not without value. However, there always comes the time where it is necessary to cast loose and use one's brain instead; for when all is said and done, it is as much by brains as by force that war is fought.[18]

Despite the critical analysis we leveled at prevailing military theories, then, we should not forget, in our haste toward the future, the lessons of history and the wisdom of our military forbears. We build our futures using what our predecessors left us. And one bit of wisdom they left is this, again from Sun Tzu: "If not in the interests of the state, do not act. If you cannot succeed, do not use troops. If you are not in danger, do not fight."[19] With that in mind, we can hope our theories go untested for a long time.

John Keegan wrote, "Knowledge, the conventional wisdom has it, is power; but knowledge cannot destroy or deflect or damage or even defy an offensive initiative by an enemy unless the possession of knowledge is also allied to objective force."[20] No commander who hopes for success comes to an encounter empty-handed; or, if literally empty-handed, he need not come to it empty-minded. Absent the perfect predictor of victory, we operate on faith, and guts, and sometimes sheer determination. It always has been that way, and it always will be.

REFERENCES

Air Force Doctrine Document 1: Air Force Basic Doctrine. Maxwell Air Force Base, AL: United States Air Force, 17 November 2003. On-line, 14 June 2011, https://www.doctrine.af.mil/. (Superseded by the 14 October 2011 version.)

Air Force Doctrine Document 1: Air Force Basic Doctrine. Washington, DC: US Air Force, September 1997. (Superseded by the 17 November 2003 version.)

Air Force Doctrine Document 1: Air Force Basic Doctrine, Organization, and Command. Maxwell Air Force Base, AL: United States Air Force, 14 October 2011. On-line, 21 March 2012, http://www.au.af.mil/au/lemay/main.htm. (Superseded by the 27 February 2015 version.)

Air Force Doctrine Document 1-2, Air Force Glossary. Maxwell AFB: Air Force Doctrine Center, 9 July 1999. On-line, 20 July 2003, https://www.doctrine.af.mil/.

Air Force Doctrine Volume 1, Basic Doctrine. Maxwell Air Force Base, AL: United States Air Force, 27 February 2015. On-line, 5 March 2018, http://www.doctrine.af.mil/Core-Doctrine/Vol-1-Basic-Doctrine/.

Air Force Manual 1-1, Basic Aerospace Doctrine of the United States Air Force, Volume II. Maxwell Air Force Base, AL: United States Air Force, March 1992.

Aldrich, Major Richard W. "The International Legal Implications of Information Warfare." *Airpower Journal.* Volume X, Number 3, Fall 1996, pp. 99-110.

Alger, John I. *The Quest for Victory: The History of the Principles of War.* Westport, CT: Greenwood Press, 1982. Originally published in 1890.

Arkin, William. "Planning an Iraqi War But Not an Outcome." *Los Angeles Times,* 5 May 2002, p. M-1.

Army Doctrine Publication 1, The Army. Washington, DC: Department of the Army, 17 September 2012. Superseded FM-1, 14 June 2005. On-line, 2 March 2018, https://armypubs.army.mil/ProductMaps/PubForm/ADP.aspx.

Army Doctrine Publication 3-0, Operations. Washington, DC: Department of the Army, 6 October 2017. Superseded FM-3, 27 February 2008. On-line, 2 March 2018, https://armypubs.army.mil/ProductMaps/PubForm/ADP.aspx.

Arquilla, John, and David Ronfeldt. *The Advent of Netwar.* Santa Monica, CA: RAND National Defense Research Institute, 1996. On-line, 15 April 2004, http://www.rand.org/publications/MR/MR789/.

Baner, Major Carl. "Defining Air and Space Power." *Aerospace Power Chronicles.* http://www.airpower.maxwell.af.mil, 11 March 1999.

Barco, Captain Charles T. "Valuing Leadership in an Era of Prophets, Politicians, and Pugilists." *Airpower Journal.* Volume VIII, Number 3, Fall 1994, pp. 4-13.

Barnett, Jeffery R. *Future War: An Assessment of Aerospace Campaigns in 2010.* Maxwell Air Force Base, AL: Air University Press, January 1996. On-line, 18 November 2004, http://www.maxwell.af.mil/au/aul/aupress/.

Barry, Major General (select) John L., and Colonel Darrell L. Herriges. "Aerospace Integration, Not Separation." *Aerospace Power Journal.* Volume XIV, Number 2, Summer 2000, pp. 42-47.

Belote, Major Howard D. "The Weaponization of Space—It Doesn't Happen in a Vacuum." *Aerospace Power Journal.* Volume XIV, Number 1, Spring 2000, pp. 46-52.

Builder, Carl H. *The Icarus Syndrome: The Role of Air Power Theory in the Evolution and Fate of the US Air Force.* New Brunswick, NJ: Transaction Publishers, 1994.

Bush, George W. "Remarks at a Ceremony Honoring Seven Nations on Accession to the North Atlantic Treaty Organization," 29 March 2004. *Weekly Compilation of Presidential Documents*, Volume 40, Number 14. On-line, 27 July 2004, http://www.gpoaccess.gov/wcomp/index.html.

Carlson, Lieutenant General Bruce. "Protecting Global Utilities: Safeguarding the Next Millennium's Space-Based Public Services." *Aerospace Power Journal*. Volume XIV, Number 2, Summer 2000, pp. 37-41.

Clausewitz, Carl von. *On War*. Edited and Translated by Michael Howard and Peter Paret. Princeton, NJ: Princeton University Press, 1976.

Columbia *Accident Investigation Board Report Volume 1*. Washington, DC: National Aeronautics and Space Administration, August 2003.

Correll, John T. "The Military-Civilian Gap." *Air Force Magazine*. Volume 83, Number 1, January 2000, p. 2.

Craig, Gordon A. "Delbruck: The Military Historian." In Peter Paret, Editor, *Makers of Modern Strategy: from Machiavelli to the Nuclear Age*. Princeton: Princeton University Press, 1986.

Crowl, Philip A. "Alfred Thayer Mahan: The Naval Historian," in Peter Paret, Editor, *Makers of Modern Strategy: from Machiavelli to the Nuclear Age*. Princeton: Princeton University Press, 1986.

Daalder, Ivo H., and Michael E. O'Hanlon. "Unlearning the Lessons of Kosovo." *Foreign Policy*. Number 116, Fall 1999, pp. 128-40.

Deming, W. Edwards. *The New Economics for Industry, Government, Education*. Cambridge, Massachusetts: MIT Center for Advanced Engineering Study, 1993.

Diamond, Jared. *Guns, Germs, and Steel*. New York: W.W. Norton and Company, 1998.

DiNardo, R.L., and Daniel J. Hughes. "Some Cautionary Thoughts on Information Warfare." *Airpower Journal*. Volume IX, Number 4, Winter 1995, pp. 69-79.

Dinerman, Taylor. "'Near Space:' a new area of operations or a new Pentagon buzzword?" *The Space Review*. 20 September 2004. On-line, 13 April 2005, http://www.thespacereview.com/article/230/1.

Dippold, Major Mark K., USAF. "Air Occupation: Asking the Right Questions." *Airpower Journal*. Volume XI, Number 4, Winter 1997, pp. 69-84.

DOD Dictionary of Military and Associated Terms. Washington,

DC: US Department of Defense, February 2018. On-line, 26 February 2018, http://www.jcs.mil/Doctrine/.

Drew, Dennis M. "Inventing A Doctrine Process." *Airpower Journal.* Volume IX, Number 4, Winter 1995, pp. 42-52.

Durant, Will and Ariel. *The Story of Civilization, Vol. 10: Rousseau and Revolution.* New York: MJF Books, 1967.

Eccles, Henry E. *Military Concepts and Philosophy.* New Brunswick, NJ: Rutgers University Press, 1965.

Echevarria, Antulio J. II. *Toward an American Way of War.* Carlisle, PA: US Army War College Strategic Studies Institute, March 2004.

"Exit strategy: How best to get out." *E-CYCLOPEDIA: The words behind the headlines.* On-line, 20 October 2005, http://news.bbc.co.uk/1/hi/magazine/3721983.stm.

Field Manual 1: The Army. Washington, DC: United States Army, 14 June 2005. On-line, 4 July 2011, http://www.army.mil/fm1/.

Field Manual 3-0: Operations. Washington, DC: United States Army, 27 February 2008. On-line, 4 July 2011, http://downloads.army.mil/fm3-0/FM3-0.pdf.

Field Manual 3-0, Operations (Change 1). Washington, DC: Department of the Army, 6 December 2017. On-line, 10 May 2018, https://armypubs.army.mil/ProductMaps/PubForm/FM.aspx.

Forsyth, Major Michael, "Finesse: A Short Theory of War." *Military Review,* Volume 84, Issue 4, July-August 2004, pp. 17-9.

Forward, Robert L. "Surgical Strikes from Space Using Solar Sail Statites." In Robert L. Forward, *21st Century Space Propulsion Study Addendum.* Phillips Laboratory Propulsion Directorate Technical Report PL-TR-91-3022, June 1991.

Frost, Robert S. *The Growing Imperative to Adopt "Flexibility" as an American Principle of War.* Carlisle, PA: Army War College Strategic Studies Institute, 15 October 1999.

Fukuyama, Francis. *The End of History and the Last Man.* New York: Perennial, 2002. First published in 1993

Grandin, Temple, and Catherine Johnson. *Animals in Translation: Using the Mysteries of Autism to Decode Animal Behavior.* New York: Scribner, 2005.

Gray, Colin S. *Modern Strategy.* Oxford: Oxford University Press, 1999.

Gray, Colin S., and John B. Sheldon. "Space Power and the

Revolution in Military Affairs—A Glass Half Full?" *Airpower Journal*. Volume XIII, Number 3, Fall 1999, pp. 23-38.

Griess, Thomas E. "A Perspective on Military History." In *A Guide to the Study and Use of Military History*. US Army Center of Military History, 1982 reprint.

Gunzinger, Lieutenant Colonel Mark A. "Airpower as a Second Front." *Airpower Journal*. Volume IX, Number 3, Fall 1995, pp. 63-73.

Hart, Gary. *The Fourth Power: A Grand Strategy for the United States in the Twenty-First Century*. Oxford: Oxford University Press, 2004.

Hinen, Colonel Anthony L. USAF. "Kosovo: 'The Limits of Air Power II.'" *Air and Space Power Chronicles*. 16 May 2002. On-line, 22 April 2004, http://www.airpower.maxwell.af.mil/airchronicles/cc/hinen.html.

Holley, I.B. Jr. "A Modest Proposal: Making Doctrine More Memorable." *Airpower Journal*. Volume IX, Number 4, Winter 1995, pp. 14-20.

Holley, I.B. Jr. *Technology and Military Doctrine: Essays on a Challenging Relationship*. Maxwell AFB, Alabama: Air University Press, August 2004.

Holmes, Richard. "Battle: The Experience of Modern Combat." In Charles Townshend, editor, *The Oxford History of Modern War*. Oxford: Oxford University Press, 2000.

Jensen, Colonel Owen. "Information Warfare: Principles of Third-Wave War." *Airpower Journal*. Volume VIII, Number 4, Winter 1994, pp. 35-43.

Joint Publication 0-2, Unified Action Armed Forces (UNAAF). Washington, DC: US Department of Defense, 10 July 2001. On-line, 12 April 2004, http://www.dtic.mil/doctrine/s_index.html.

Joint Publication 1, Joint Warfare of the Armed Forces of the United States. Washington, DC: US Department of Defense, 14 November 2000. On-line, 12 April 2004, http://www.dtic.mil/doctrine/s_index.html. (Superseded by the 2 May 2007 version.)

Joint Publication 1, Doctrine for the Armed Forces of the United States. Washington, DC: US Department of Defense, 2 May 2007, Change 1 - 20 March 2009. On-line, 14 June 2011, http://www.dtic.mil/doctrine/s_index.html.

Joint Publication 1, Doctrine for the Armed Forces of the United States. Washington, DC: US Department of Defense, 25 March 2013;

incorporating Change 1, 12 July 2017. On-line, 7 March 2018, http://www.jcs.mil/Doctrine/Joint-Doctrine-Pubs/Capstone-Series/.

Joint Publication 1-02, Department of Defense Dictionary of Military and Associated Terms. Washington, DC: US Department of Defense, 8 November 2010 (as amended through 15 May 2011). On-line, 14 June 2011, http://www.dtic.mil/doctrine/s_index.html.

Joint Publication 3-0, Doctrine for Joint Operations. Washington, DC: US Department of Defense, 17 September 2006, Incorporating Change 2, 22 March 2010. On-line, 14 June 2011, http://www.dtic.mil/doctrine/s_index.html.

Joint Publication 3-0, Joint Operations. Washington, DC: US Department of Defense, 17 January 2017. On-line, 7 March 2018, http://www.jcs.mil/Doctrine/Joint-Doctrine-Pubs/3-0-Operations-Series/.

Joint Publication 3-07, Joint Doctrine for Military Operations Other Than War. Washington, DC: US Department of Defense, 16 June 1995. On-line, 28 February 2002, http://www.dtic.mil/doctrine/s_index.html.

Joint Publication 5-0, Doctrine for Planning Joint Operations. 13 April 1995. On-line, 12 October 2005, http://www.dtic.mil/doctrine/s_index.html.

Joint Publication 5-0, Joint Operation Planning. Washington, DC: US Department of Defense, 26 December 2006. On-line, 14 June 2011, http://www.dtic.mil/doctrine/s_index.html.

Jomini, Antoine Henri. *The Art of War.* Translated by G.H. Mendell and W.P. Craighill. Westport, CT: Greenwood Press, Publishers, undated.

Keegan, John. *A History of Warfare.* New York: Alfred A. Knopf, 1993.

Keegan, John. *Intelligence in War: Knowledge of the Enemy from Napoleon to Al-Qaeda.* New York: Alfred A. Knopf, 2003.

Keegan, John. *The Face of Battle.* New York: Viking Press, 1976.

Keegan, John. *The Mask of Command.* New York: Penguin Books, 1987.

Keegan, John. *The Second World War.* New York: Viking Penguin, 1989.

Kenney, George C. *General Kenney Reports: A Personal History of the Pacific War.* Washington, DC: Reprinted by the Office of Air Force History, 1987.

Knowles, Billy M. "Hold, Withdraw, or Advance: The Role of Tactical Airlift in the Equation." *Airpower Journal*. Volume III, Number 2, Spring 1989, pp. 41-7.

Knowles, David. *The Evolution of Medieval Thought*. Second Edition. London: Longman Group, 1988.

Kosovo/Operation Allied Force After-Action Report. Washington, DC: US Department of Defense, 31 January 2000. On-line, 11 October 2005, www.defenselink.mil.

Kratman, Tom. "Indirectly Mistaken Decision Cycles." Baen.com, 2012. On-line, 20 June 2018, https://www.baen.com/decisioncycles.

Lehman, John. "USS Cole: An Act of War." *Washington Post*, 15 October 2000, p. B-7. On-line, 5 October 2005, http://www.washingtonpost.com.

Lewis, Clarence Irving. *Mind and the World Order: Outline of a Theory of Knowledge*. New York: Dover Publications, Inc., 1956.

Liddell Hart, B.H. *History of the Second World War*. New York: Perigee Books, 1982. (Original copyright 1970.)

Liddell Hart, B.H. *Strategy*, Second Revised Edition. New York: Signet, 1967.

Lind, William S., "John Boyd's Art of War." *The American Conservative*, 16 August 2013. Online, 26 February 2018, http://www.theamericanconservative.com/articles/john-boyds-art-of-war/.

Lind, William S., and LtCol Gregory A. Thiele, USMC, *4th General Warfare Handbook*. Kouvola, Finland: Castalia House, 2015.

Mahan, Alfred Thayer. *The Influence of Sea Power Upon History, 1660-1783*. Boston: Little, Brown, and Company, 1903 reprint.

Mandelbaum, Michael. "A Perfect Failure: NATO's War in Yugoslavia." *Foreign Affairs*, September/October 1999, Volume 78, Number 5, pp. 2-8.

Mann, Edward C. III. *Thunder and Lightning: Desert Storm and the Airpower Debates*. Maxwell AFB, AL: Air University Press, 1995.

Mann, Lieutenant Colonel Edward. "One Target, One Bomb: Is the Principle of Mass Dead?" *Airpower Journal*. Volume VII, Number 1, Spring 1993, pp. 35-43.

Mantz, Michael R. *The New Sword: A Theory of Space Combat Power*. Maxwell AFB, AL: Air University Press, 1995.

Marine Corps Doctrine Publication (MCDP) 1, Warfighting. Washington, DC: Department of the Navy, 20 June 1997. On-line, 30

March 2018, http://www.marines.mil/News/Publications/MCPEL/Electronic-Library-Display/Article/899837/mcdp-1/.

McKinley, Lieutenant Colonel Cynthia A.S. "The Guardians of Space—Organizing America's Space Assets for the Twenty-First Century." *Aerospace Power Journal*. Volume XIV, Number 1, Spring 2000, pp. 37-45.

McNeilly, Mark. *Sun Tzu and the Art of Modern Warfare*. Oxford: Oxford University Press, 2001.

Meilinger, Phillip S. *Ten Propositions Regarding Air Power*. Air Force History and Museums Program, 1995.

Metz, Steven. "The Operational Level of Nuclear War Fighting: Missing or Unnecessary?" *Airpower Journal*. Volume VII, Number 1, Spring 1993, pp. 13-20.

Meyer, Milton W. *Japan: A Concise History*. Third Edition. Lanham, MD: Littlefield Adams Quality Paperbacks, 1993.

Millsap Ralph, and D.B. Posey, "Organizational Options for the Future Aerospace Force." *Aerospace Power Journal*. Volume XIV, Number 2, Summer 2000, pp. 48-52.

Moen, Ronald D., Thomas W. Nolan, and Lloyd P. Provost. *Improving Quality Through Planned Experimentation*. New York: McGraw-Hill, 1991.

Moseley, Alexander. *A Philosophy of War*. New York: Algora Publishing, 2002.

Mowbray, James A. "Air Force Doctrine Problems: 1926-Present." *Airpower Journal*. Volume IX, Number 4, Winter 1995, pp. 21-41.

Musashi, Miyamoto. *Go Rin No Sho (A Book of Five Rings)*. Translated by Victor Harris. Woodstock, NY: Overlook Press, 1974.

"Near-space demo aims to prove boost in combat communication ability." Air Force Space Command Press Release. Colorado Springs, CO: Air Force Space Command News Service, 8 March 2005. On-line, 11 May 2005, http://www.peterson.af.mil/hqafspc/News/News_Asp/nws_tmp.asp?storyid=05-070.

O'Connell, Robert L. *Of Arms and Men: A History of War, Weapons, and Aggression*. Oxford: Oxford University Press, 1989.

Paret, Peter. "Clausewitz," in Peter Paret, Editor, *Makers of Modern Strategy: from Machiavelli to the Nuclear Age*. Princeton: Princeton University Press, 1986.

Pellegrini, Robert P. *The Links Between Science, Philosophy and*

Military Theory: Understanding the Past, Implications for the Future. Maxwell Air Force Base, AL: Air University Press, August 1997.

Pintner, Walter. "Russian Military Thought: The Western Model and the Shadow of Suvorov." In Peter Paret, Editor, *Makers of Modern Strategy: from Machiavelli to the Nuclear Age*. Princeton: Princeton University Press, 1986.

Porch, Douglas. "Imperial Wars: From the Seven Years War to the First World War." In Charles Townshend, editor, *The Oxford History of Modern War*. Oxford: Oxford University Press, 2000.

Renatus, Flavius Vegetius. *Epitoma Rei Militaris*. Edited with an English translation by Leo F. Stelton. New York: Peter Lang, 1990.

Report of the Commission to Assess United States National Security Space Management and Organization (Washington, DC: January 11, 2001). On-line, 3 July 2011, http://www.dod.gov/pubs/space20010111.html.

Rife, Major Shawn P. "On Space-Power Separatism." *Airpower Journal*. Volume XIII, Number 1, Spring 1999, pp. 21-31.

Rinaldi, Major Steven M. *Beyond the Industrial Web: Economic Synergies and Targeting Methodologies*. Maxwell AFB, AL: Air University Press, April 1995. On-line, 13 October 2005, http://www.maxwell.af.mil/au/aul/aupress/.

Rinehart, Captain Graham W. "A New Paradigm for Organizational Structure." *Airpower Journal*. Volume VI, Number 1, Spring 1992, pp. 43-53.

Rinehart, Gray (editor), Erik Eliasen, Steve Julian, Mike Lutton, Keith Phillips, John Taylor, Mark VanVoorhis, and James H. Galt-Brown. "Toward Space War." *High Frontier*, Volume 1, Number 3, Winter 2005, pp. 47-53.

Rinehart, Gray. *A Long Enough Lever—Propulsion and Power for U.S. Superiority in Space*. Unpublished.

Rinehart, Gray. *Quality Education: Applying the Philosophy of Dr. W. Edwards Deming to Transform the Educational System*. Milwaukee, WI: ASQC-Quality Press, 1993.

Ringo, John, and Tom Kratman. *Watch on the Rhine—Die Wacht am Rhein*. Riverdale, New York: Baen Books, 2005.

Schmitt, Eric. "U.S. Plan for Iraq is Said to Include Attack on 3 Sides." *New York Times*, 5 July 2002, p. A-1.

Schwarzkopf, H. Norman, with Peter Petre. *It Doesn't Take a Hero*. New York: Linda Grey Bantam Books, 1992.

Shakespeare, William. *Hamlet*.

Shy, John. "Jomini," in Peter Paret, Editor, *Makers of Modern Strategy: from Machiavelli to the Nuclear Age*. Princeton: Princeton University Press, 1986.

Skelton, Representative Ike. "It Ain't New." *Aerospace Power Journal*. Volume XIV, Number 3, Fall 2000, pp. 5-12.

Smith, Major Russell J. "Developing an Air Campaign Strategy." *Aerospace Power Chronicles*. http://www.airpower.maxwell.af.mil, 23 November 1999.

Smith, Rupert. *The Utility of Force—The Art of War in the Modern World*. London: Allen Lane, 2005.

Sontag, Sherry, Christopher Drew, and Annette Lawrence Drew. *Blind Mans Bluff: The Untold Story of American Submarine Espionage*. New York: Public Affairs, 1998.

"Space Policy." *Department of Defense Directive* 3100.10, 9 July 1999.

"Space Policy." *Department of Defense Directive* 3100.10, 18 October 2012. Incorporating Change 1, 4 November 2016. On-line, 30 April 2018, *https://www.hsdl.org/?abstract&did=796643*.

Sun Tzu. *The Art of War*. Translated by Samuel B. Griffith. London: Oxford University Press, 1963.

Sun Tzu. *The Art of War*. Translated by Thomas Cleary. Boston: Shambhala, 1988.

Szafranski, Richard, and Martin C. Libicki. "'. . . Or Go Down in Flame?' Toward an Airpower Manifesto for the Twenty-First Century." *Airpower Journal*. Volume X, Number 3, Fall 1996, pp. 65-77.

Szafranski, Richard. "The Problems With Bees and Bombs." *Airpower Journal*. Volume IX, Number 4, Winter 1995, pp. 94-8.

The Commission on the Intelligence Capabilities of the United States Regarding Weapons of Mass Destruction, 31 March 2005. On-line, 30 April 2005; http://www.wmd.gov/report/.

The North Atlantic Treaty, 14 April 1949. On-line, 7 October 2005, http://www.nato.int/docu/basictxt/treaty.htm.

Thomas, Evan, Rod Nordland, and Christian Caryl. "Operation Hearts and Minds." *Newsweek*. Volume 143, Number 1, 29 December 2003-5 January 2004, pp. 34-43.

Townshend, Charles. "The Shape of Modern War," in Charles Townshend, editor, *The Oxford History of Modern War*. Oxford: Oxford University Press, 2000.

Tsouras, Peter G., Editor. *Warriors' Words: A Quotation Book.* London: Arms and Armour Press, 1992.

Van Creveld, Martin. "Technology and War II: Postmodern War?" In Charles Townshend, editor, *The Oxford History of Modern War.* Oxford: Oxford University Press, 2000.

Van Creveld, Martin. *The Transformation of War.* New York: The Free Press, 1991.

Vasquez, John A. *The War Puzzle.* Cambridge: Cambridge University Press, 1993.

Vegetius: Epitome of Military Science. Translated with notes and introduction by N.P. Milner. Liverpool: Liverpool University Press, 1993.

Vesely, W.E., F.F. Goldberg, N.H. Roberts, and D.F. Haasl. *Fault Tree Handbook,* NUREG-0492. Washington, DC: US Nuclear Regulatory Commission, January 1981.

Votaw, John F. "An Approach to the Study of Military History." In *A Guide to the Study and Use of Military History.* US Army Center of Military History, 1982 reprint.

Warden, John A. *The Air Campaign: Planning for Combat.* Washington: Pergamon-Brassey's, 1989.

Warden, Colonel John A. III. "The Enemy as a System." *Airpower Journal.* Volume IX, Number 1, Spring 1995, pp. 40-55.

Ware, Lewis. "Some Observations of the Enemy As a System." *Airpower Journal.* Volume IX, Number 4, Winter 1995, pp. 87-93.

Weigley, Russell F. *The American Way of War: A History of United States Military Strategy and Policy.* New York: Macmillan Publishing Company, Inc., 1973.

Wheeler, Donald J. *Understanding Industrial Experimentation.* 2nd Edition. Knoxville, TN: SPC Press, 1988.

NOTES

1. War and Theory

1. Clarence Irving Lewis, *Mind and the World Order: Outline of a Theory of Knowledge* (New York: Dover Publications, Inc., 1956), p. 3.
2. See, interestingly, Mark McNeilly, *Sun Tzu and the Art of Modern Warfare* (Oxford: Oxford University Press, 2001), pp. 182ff.

   "Interestingly" probably only to the author, because the line about comparisons between Clausewitz and Sun Tzu originated in the author's 1998 article that spawned this entire text. Finding that someone had attempted an in-depth comparison was a pleasant surprise.
3. A retired US Navy Rear Admiral writes, "Too many military men tend to be contemptuous of theory and history." Henry E. Eccles, *Military Concepts and Philosophy* (New Brunswick, NJ: Rutgers University Press, 1965), p. 24.
4. B.H. Liddell Hart, *Strategy*, Second Revised Edition (New York: Signet, 1967), p. 3. (Emphasis added.)
5. *Air Force Doctrine Volume 1, Basic Doctrine* (Maxwell Air Force Base, AL: United States Air Force, 27 February 2015). On-line, 5 March 2018, http://www.doctrine.af.mil/Core-Doctrine/Vol-1-Basic-Doctrine/. (No page numbers in online document.)

   The quote on experience is unchanged from the previous *Air Force Doctrine Document 1: Air Force Basic Doctrine, Organization, and Command* (Maxwell Air Force Base, AL: United States Air Force, 14 October 2011), p. 7. On-line, 21 March 2012, http://www.au.af.mil/au/lemay/main.htm.
6. *Joint Publication 1, Doctrine for the Armed Forces of the United States* (Washington, DC: US Department of Defense, 25 March 2013; incorporating Change 1, 12 July 2017), p. I-3. On-line, 7 March 2018, http://www.jcs.mil/Doctrine/Joint-Doctrine-Pubs/Capstone-Series/.

   The only other mention of "theory" is in Appendix B, p. B-1: "Every joint leader is expected to be a student of the art and science of war. Officers especially are expected to have a solid foundation in military theory and philosophy, and knowledge of military history and the timeless lessons to be gained from it."
7. *Joint Publication 3-0, Joint Operations* (Washington, DC: US Department of Defense, 17 January 2017), p. I-1. On-line, 7 March 2018, http://www.jcs.mil/Doctrine/Joint-Doctrine-Pubs/3-0-Operations-Series/.

   This is somewhat different from a previous version, which based doctrine on "a sound base of warfighting theory and practical experience." *Joint Publication 3-0, Doctrine for Joint Operations* (Washington, DC: US Department of Defense, 17 September 2006, Incorporating Change 2, 22 March 2010), p. II-1. On-line, 14 June 2011, http://www.dtic.mil/doctrine/s_index.html.
8. *Field Manual 1: The Army* (Washington, DC: United States Army, 14 June 2005), p. 1-21. On-line, 4 July 2011, http://www.army.mil/fm1/. Contrast with *Army Doctrine Publication 1, The Army* (Washington, DC: Department of the Army, 17 September 2012). Superseded FM-1, 14 June 2005. On-line, 2 March 2018, https://armypubs.army.mil/ProductMaps/PubForm/ADP.aspx.

   Likewise, the old FM 3-0 (formerly FM 100-5) mentioned "theory" only once, also: in the Preface, describing its own Chapter 6. *Field Manual 3-0: Operations* (Washington, DC: United States Army, 27 February 2008), p. v. On-line, 4 July 2011, http://downloads.army.mil/fm3-0/FM3-0.pdf. Contrast with *Army Doctrine Publication 3-0, Operations* (Washington, DC: Department of the Army, 6 October

2017). Superseded FM-3, 27 February 2008. On-line, 2 March 2018, https://armypubs.army.mil/ProductMaps/PubForm/ADP.aspx.

9. "Principles of War," in *Air Force Manual 1-1, Basic Aerospace Doctrine of the United States Air Force, Volume II* (Maxwell Air Force Base, AL: United States Air Force, March 1992), p. 14.
10. *Air Force Doctrine Document 1* (27 February 2015).
11. *Air Force Doctrine Document 1* (27 February 2015).

    The phrasing is almost the same as used in previous versions as well. See the 2011 document, as well as *Air Force Doctrine Document 1: Air Force Basic Doctrine* (Maxwell Air Force Base, AL: United States Air Force, 17 November 2003), p. 20. On-line, 14 June 2011, *https://www.doctrine.af.mil/*.
12. The Navy's doctrine library is, as of early 2018, maintained within a system reserved to US government "Common Access Card" holders.
13. *Marine Corps Doctrine Publication (MCDP) 1, Warfighting* (Washington, DC: Department of the Navy, 20 June 1997), p. 57. On-line, 30 March 2018, http://www.marines.mil/News/Publications/MCPEL/Electronic-Library-Display/Article/899837/mcdp-1/.
14. *MCDP-1, Warfighting*, p. 64.
15. Use of the masculine pronoun is not meant to marginalize or exclude female warriors; rather, it is a convenience that avoids more cumbersome constructions.

2. War in the Abstract

1. Quoted in Peter G. Tsouras, Editor, *Warriors' Words: A Quotation Book* (London: Arms and Armour Press, 1992), p. 465. The context of the quote is instructive, as well:

    "War is at best barbarism . . . . Its glory is all moonshine. It is only those who have neither fired a shot nor heard the shrieks and groans of the wounded, who cry aloud for blood, more vengeance, more desolation. War is hell."
2. Rupert Smith, *The Utility of Force—The Art of War in the Modern World* (London: Allen Lane, 2005), p. 1.
3. John Keegan, "Please, Mr. Blair, never take such a risk again," *London Daily Telegraph*, 6 June 1999. Quoted in Colonel Anthony L. Hinen, USAF, "Kosovo: 'The Limits of Air Power II,'" *Air and Space Power Chronicles*, 16 May 2002. On-line, 22 April 2004, *http://www.airpower.maxwell.af.mil/airchronicles/cc/hinen.html*.

    This was also discussed in "Unlearning the Lessons of Kosovo," in which the first paragraph stated, "The air campaign succeeded," but subsequent sections applied important caveats to that conclusion. Ivo H. Daalder and Michael E. O'Hanlon, "Unlearning the Lessons of Kosovo," *Foreign Policy*, Number 116, Fall 1999, p. 128; see also pp. 131-2.
4. John A. Vasquez, *The War Puzzle* (Cambridge: Cambridge University Press, 1993), p. 14.
5. Alexander Moseley, *A Philosophy of War* (New York: Algora Publishing, 2002), p. 13.
6. See Martin van Creveld, *The Transformation of War* (New York: The Free Press, 1991), especially chapters 5-7.
7. The definition continues: "the employment of armed forces against a foreign power, or against an opposing party in the state." The OED also allows for "war" to be "applied [poetically] or rhetorically to any kind of active hostility or contention between living beings, or of conflict between opposing forces or principles."
8. *DOD Dictionary of Military and Associated Terms* (Washington, DC: US Department of Defense, February 2018), p. 247. On-line, 26 February 2018, http://www.jcs.mil/Doctrine/.

    The previous version, *Joint Publication 1-02, Department of Defense Dictionary of Military and Associated Terms* (Washington, DC: US Department of Defense, 12 April 2001 [as amended through 31 August 2005]), included more definitions, for such things as "war games," but not for "war" or "warfare." On-line, 11 October 2005, *http://www.dtic.mil/doctrine/jpreferencepubs.htm.*

Similarly, there are no entries for "battle" or "combat" in the current dictionary, nor were there in the previous version.

9. Vasquez, *The War Puzzle*, p. 15.
10. See John Arquilla and David Ronfeldt, *The Advent of Netwar* (Santa Monica, CA: RAND National Defense Research Institute, 1996). On-line, 15 April 2004, *http://www.rand.org/publications/MR/MR789/.*
11. Of those listed, the IRA and PLO are probably the most well-known to US residents; aside from the KLA there are dozens if not hundreds of groups worldwide who qualify as terrorists. Of course, if the IRA lives up to its pledge to cease using violence to attain its ends, it would no longer qualify as a terrorist organization.
12. Van Creveld discusses the prevalence of low-intensity conflict throughout *The Transformation of War*. Also, see John Keegan, *A History of Warfare* (New York: Alfred A. Knopf, 1993), for a fascinating discussion of how tribal warfare in ancient times was a more stylistic affair usually fought by selected champions instead of by entire groups.
13. General Smith devotes most of the last third of *The Utility of Force* to this new type of conflict.
14. Charles Townshend, "The Shape of Modern War," in Charles Townshend, editor, *The Oxford History of Modern War* (Oxford: Oxford University Press, 2000), p. 15.

3. The Risk of War

1. Isaiah 2:4.
2. See Fukuyama, *The End of History*, especially pp. 146ff, 247ff, 311ff, and 329ff.

   "Nations may on occasion conquer their neighbors as a matter of necessity, in self-defense, or to build up population and resources for the future. But above and beyond such considerations was the desire of man to be recognized—the pleasure that a Roman general felt during his triumph when his opponent was paraded through the streets in chains to the cheers of the multitude." *The End of History*, p. 184.

   Thankfully, as Fukuyama points out, liberal democracies are not as susceptible to these motivations; however, "the true coin of the realm in international politics is military power. Other forms of power such as natural resources or industrial capability are important, but primarily as a means of creating the military capabilities for self-defense." *The End of History*, p. 249.
3. Despite recent efforts—and some might say recent progress—at including women in warfighting roles, "Warfare is . . . the one human activity from which women, with the most insignificant exceptions, have always and everywhere stood apart." Keegan, *A History of Warfare*, p. 76.
4. In the 1980s, science fiction author Dr. Jerry Pournelle first edited an anthology series entitled *There Will Be War*.
5. Keegan, *A History of Warfare*, p. 83. See also p. 384.
6. George Wilhelm Friedrich Hegel's philosophical idea of the "first man" and his need for recognition may, at least in part, explain this. See Fukuyama, *The End of History*, especially chapter 14.
7. *Air Force Doctrine Volume 1, Basic Doctrine* (Maxwell Air Force Base, AL: United States Air Force, 27 February 2015). On-line, 5 March 2018, http://www.doctrine.af.mil/Core-Doctrine/Vol-1-Basic-Doctrine/. (No page numbers in online document.)
8. Keegan, *The Second World War*, concludes with a discussion of the risk of war. Whereas in the past the development of better and better weapons led nations to consider war more attractive—because they offered advantages that increased the chances of victory—the exponential leap in destructiveness achieved with nuclear weapons seems to have checked that tendency.

See also Keegan, *The Face of Battle*, p. 319; and *A History of Warfare*, p. 56.

9. Martin van Creveld, *The Transformation of War*, pp. 20-1.
10. Echevarria, *Toward an American Way of War*, p. 5.
11. *The North Atlantic Treaty*, 14 April 1949. On-line, 7 October 2005, *http://www.nato.int/docu/basictxt/treaty.htm*. All quotes of the treaty articles are from this source.
12. The former Yugoslav republic of Slovenia joined NATO in March 2004.
13. See Ike Skelton, "It Ain't New," *Aerospace Power Journal*, Volume XIV, Number 3, Fall 2000, p. 8.
14. President Bush stated that "the Alliance proved its enduring worth by stopping ethnic cleansing in Bosnia and by ousting the armies of a tyrant in Kosovo." George W. Bush, "Remarks at a Ceremony Honoring Seven Nations on Accession to the North Atlantic Treaty Organization," 29 March 2004, *Weekly Compilation of Presidential Documents*, Volume 40, Number 14, p. 496. On-line, 27 July 2004, *http://www.gpoaccess.gov/wcomp/index.html*.
15. Keegan, *A History of Warfare*, p. 58.
16. Michael Mandelbaum, "A Perfect Failure: NATO's War in Yugoslavia," *Foreign Affairs*, September/October 1999, Volume 78, Number 5, p. 5.

4. What We Know, and How We Know It

1. Moseley, *A Philosophy of War*, p. 225.
2. James A. Mowbray, "Air Force Doctrine Problems: 1926-Present," *Airpower Journal*, Volume IX, Number 4, Winter 1995, p. 37.

   As stated by General Charles C. Krulak, USMC, "Doctrine must continue to evolve based on growing experience, advancements in theory, and the changing face of war itself." *Joint Publication 1, Joint Warfare of the Armed Forces of the United States* (Washington, DC: US Department of Defense, 14 November 2000), p. I-9. On-line, 12 April 2004, *http://www.dtic.mil/doctrine/s_index.html*.

   However, as pointed out by Captain Mike Scardera, theory ought to be essential to developing doctrine but the two generally are treated separately.

   See also I.B. Holley, Jr., "A Modest Proposal: Making Doctrine More Memorable," *Airpower Journal*, Volume IX, Number 4, Winter 1995, p. 19; and Dennis M. Drew, "Inventing A Doctrine Process," *Airpower Journal*, Volume IX, Number 4, Winter 1995, p. 45.
3. "Theories explain *why*." Carl H. Builder, *The Icarus Syndrome: The Role of Air Power Theory in the Evolution and Fate of the US Air Force* (New Brunswick, NJ: Transaction Publishers, 1994), p. 206. (Emphasis in original.)
4. Lewis, *Mind and the World Order*, p. 44.

   See also W. Edwards Deming, *The New Economics for Industry, Government, Education* (Cambridge, Massachusetts: MIT Center for Advanced Engineering Study, 1993), pp. 104-10.
5. Lewis, *Mind and the World Order*, p. 230.

   "Where Lewis uses the word *concept*, Deming substitutes the word *theory*. Those readers uncomfortable with either word but familiar with the scientific method . . . may prefer to substitute the word *hypothesis*." Gray Rinehart, *Quality Education: Applying the Philosophy of Dr. W. Edwards Deming to Transform the Educational System* (Milwaukee, WI: ASQC-Quality Press, 1993), p. 29. (Emphasis in original.)

   Likewise, John Vasquez writes, "Concepts do not just reflect or capture reality, but help make reality." *The War Puzzle*, p. 19.
6. See Temple Grandin and Catherine Johnson, *Animals in Translation: Using the Mysteries of Autism to Decode Animal Behavior* (New York: Scribner, 2005), especially Chapter 2, "How Animals Perceive the World."

   Dr. Grandin refers to the human mode of perception as "abstractification." She writes, "a human being walking through an alley won't see, much less be bothered by, sparkling puddles or shiny spots on metal or jiggling chains [that would catch an animal's attention] . . . . Normal human beings are blind to anything they're not paying attention to . . . . Instead, a normal person's brain uses the detailed raw data of the world to form a generalized concept or schema, and that's what reaches consciousness." *Animals in Translation*, pp. 50-1, 65.

7. From 1990 to 2004 the UN Security Council passed several dozen resolutions pertaining to Iraq, many of which dealt with matters such as the "Oil For Food" program. The details of even the pertinent resolutions are too involved to explore here. Some of the important resolutions dealing with the Gulf Wars include (emphasis added):

Resolution 660 (1990), 2 August 1990

- Condemned Iraqi invasion of Kuwait
- Demanded immediate and unconditional withdrawal

Resolution 661 (1990), 6 August 1990

- Imposed economic sanctions on Iraq

Resolution 662 (1990), 9 August 1990

- Voided Iraq's annexation of Kuwait

Resolution 678 (1990), 29 November 1990

- Granted "one final opportunity" for Iraq to comply with 660 (1990)
- Authorized Member States to enforce 660 (1990), et al, by "all necessary means"

Resolution 686 (1991), 2 March 1991

- Took note of Iraq's "agreement to comply fully"with previous resolutions
- Demanded "Iraq implement its acceptance"
- Demanded Iraq "cease hostile or provocative actions by its forces against all Member States, including missile attacks and flights of combat aircraft"
- Demanded Iraq "provide all information and assistance in identifying . . . any chemical and biological weapons in Kuwait [and] areas of Iraq where forces of Member States . . . are present temporarily"

Resolution 687 (1991), 3 April 1991

- Expressed concern over "reports in the hands of Member States that Iraq has attempted to acquire materials for a nuclear-weapons programme" and deplored "threats made by Iraq during the recent conflict to make use of terrorism"
- Decided that "Iraq shall unconditionally accept the destruction, removal, or rendering harmless, under international supervision, of . . . all chemical and biological weapons [and] all ballistic missiles" with range >150 km
- Decided to form a Special Commission to inspect and destroy Iraq's chemical, biological, and missile programs
- Decided on a similar regime of inspection and destruction for Iraq's nuclear weapons, under the International Atomic Energy Agency (IAEA)
- Required "Iraq to inform the Council that it will not commit or support any act of international terrorism or allow any [terrorist] organization . . . to operate within its territory"

Resolution 688 (1991), 5 April 1991

- Condemned Iraq's repression of its citizens

Resolution 699 (1991), 17 June 1991

- Confirmed the Special Commission and IAEA's authority to destroy, remove, or render harmless Iraq's chemical, biological, and nuclear weapons under resolution 687 (1991) "until such activities have been completed"

Resolution 707 (1991), 15 August 1991

- Condemned "Iraq's serious violation . . . which constitutes a material breach" of resolution 687 (1991)
- Demanded "without further delay full, final and complete disclosure" of weapons and missile programs, and that Iraq "cease immediately any attempt to conceal, move or destroy any [related] material or equipment"and "halt all nuclear activities of any kind"

Resolution 715 (1991), 11 October 1991

- Demanded "that Iraq meet unconditionally all its obligations . . . and cooperate fully with the Special Commission" and IAEA

Resolution 1060 (1996), 12 June 1996

- Deplored Iraqi refusal "to allow access to sites designated by the Special Commission . . . a clear violation of" previous resolutions

- Demanded full cooperation with the Special Commission, viz., "immediate, unconditional and unrestricted access to any and all areas, facilities, equipment, records and means of transportation which they wish to inspect"

Resolution 1115 (1997), 21 June 1997

- Condemned Iraq's "repeated refusal . . . to allow access . . . , which constitutes a clear and flagrant violation of . . . Security Council resolutions"

- Demanded "that Iraq cooperate fully"

Resolution 1134 (1997), 23 October 1997

- Condemned Iraq's "repeated refusal . . . to allow access to sites . . . , and especially Iraqi actions endangering the safety of Special Commission personnel, the removal and destruction of documents of interest . . . and interference with the freedom of movement of Special Commission personnel"

- Decided "that such refusals to cooperate constitute a flagrant violation of Security Council resolutions"

- Demanded full cooperation

Resolution 1137 (1997), 12 November 1997

- Condemned "continued violations by Iraq of its obligations under the relevant resolutions to cooperate fully and unconditionally"

- Demanded "that Iraq cooperate fully and immediately and without conditions or restrictions with the Special Commission in accordance with the relevant resolutions"

- Restricted the travel (entry/exit) of Iraqi officials

- Expressed "the firm intention to take further measures as may be required for the implementation of this resolution"

Resolution 1154 (1998), 2 March 1998

- Commended the Secretary-General's initiative "to secure commitments from the Government of Iraq on compliance with its obligations"

- Stressed the need for "compliance by the Government of Iraq with its obligations [and] that any violation would have severest consequences"

Resolution 1194 (1998), 9 September 1998

- Condemned "the decision by Iraq of 5 August 1998 to suspend cooperation with the Special Commission and the IAEA, which constitutes a totally unacceptable contravention of its obligations under resolutions 687 (1991), 707 (1991), 715 (1991), 1060 (1996), 1115 (1997) and 1154 (1998)"

- Demanded "that Iraq rescind its above-mentioned decision and cooperate fully with the Special Commission and the IAEA"

Resolution 1205 (1998), 5 November 1998

- Condemned "the decision by Iraq of 31 October 1998 to cease cooperation with the Special Commission as a flagrant violation of resolution 687 (1991) and other relevant resolutions"

- Demanded "that Iraq rescind immediately and unconditionally the decision of 31 October 1998, as well as the decision of 5

August 1998, to suspend cooperation . . . , and that Iraq provide immediate, complete and unconditional cooperation"

Resolution 1284 (1999), 17 December 1999

- Acknowledged "progress made by Iraq towards compliance with the provisions of resolution 687 (1991)"

- Decided "to establish, as a subsidiary body of the Council, the United Nations Monitoring, Verification and Inspection Commission (UNMOVIC) which replaces the Special Commission"

- Decided "that UNMOVIC will establish and operate . . . a reinforced system of ongoing monitoring and verification"

Resolution 1441 (2002), 8 November 2002

- Decided "that Iraq has been and remains in material breach of its obligations under relevant resolutions . . . , in particular through Iraq's failure to cooperate with United Nations inspectors and the IAEA"

- Decided "to afford Iraq, by this resolution, a final opportunity to comply with its disarmament obligations under relevant resolutions of the Council; and accordingly decides to set up an enhanced inspection regime with the aim of bringing to full and verified completion the disarmament process"

- Decided "that . . . the Government of Iraq shall provide . . . , not later than 30 days from the date of this resolution, a currently accurate, full, and complete declaration of all aspects of its programmes to develop chemical, biological, and nuclear weapons, ballistic missiles, and other delivery systems"

- Decided "that false statements or omissions in the declarations . . . shall constitute a further material breach of Iraq's obligations"

- Decided "that Iraq shall provide . . . immediate, unimpeded, unconditional, and unrestricted access"

- Recalled "that the Council has repeatedly warned Iraq that it will face serious consequences as a result of its continued violations of its obligations"

The full text of UN Security Council resolutions is available at http://www.un.org/Docs/sc/unsc_resolutions.html.

8. Deming, *The New Economics*, p. 105.
9. Robert P. Pellegrini, *The Links Between Science, Philosophy and Military Theory: Understanding the Past, Implications for the Future* (Maxwell Air Force Base, AL: Air University Press, August 1997), p. 3.
10. Pellegrini, *The Links Between Science, Philosophy and Military Theory*, p. 4.
11. Lewis, *Mind and the World Order*, p. 2. (Emphasis in original.)
12. Lewis, *Mind and the World Order*, p. 346.
13. Grandin and Johnson, *Animals in Translation*, pp. 200-6.
14. Lewis, *Mind and the World Order*, p. 54.
15. Clausewitz, *On War*, p. 79.
16. Lewis, *Mind and the World Order*, p. 195.
17. Eccles, *Military Concepts and Philosophy*, p. 26. Rear Admiral Eccles also wrote on p. vi: ". . .while theory does not solve problems nor make decisions, it clarifies, explains, illuminates; and in important matters it may contribute greatly to fundamental intellectual harmony while still allowing room for great diversity of ideas and action."
18. Jomini, *The Art of War*, p. 295.
19. Lewis, *Mind and the World Order*, p. 54. (Emphasis added.)

5. Theory is the Key to Understanding

1. Deming, *The New Economics*, p. 107.
2. As Lewis writes, "even that minimum of cognition which consists in naming is an interpretation which implicitly asserts certain relations between the given and further experience," and "it should always have been clear . . . that the application of concepts in naming and recognizing objects, itself implies characters of the object which are not now presented but wait upon further experience to be revealed." Lewis, *Mind and the World Order*, pp. 132 and 294.
3. "As with metaphysical theories, most people do not consider epistemology to any great extent, but that does not mean they do not possess some rudimentary epistemological theories that govern their attempts to gain knowledge." Moseley, *A Philosophy of War*, p. 225.
4. Eccles, *Military Concepts and Philosophy*, p. 27.
5. Holley, *Technology and Military Doctrine*, p. 95.

6. DiNardo and Hughes, "Some Cautionary Thoughts on Information Warfare," p. 77.
7. Lewis, *Mind and the World Order*, p. 385.
8. Lewis, *Mind and the World Order*, p. 195.
9. LTC Tom Kratman, USA, personal communication, 20 October 2018.

6. Theory is Like a Good Attitude

1. "One of the drawbacks of history is that we cannot rerun the episode or the battle in the same way we can rerun a scientific experiment in the laboratory." I.B. Holley, Jr., *Technology and Military Doctrine: Essays on a Challenging Relationship* (Maxwell AFB, Alabama: Air University Press, August 2004), p. 95.
2. Musashi, *Go Rin No Sho*, p. 44.
3. Clausewitz, *On War*, p. 102. (Emphasis in original.)

   Clausewitz continued, "*Coup d'oeil* therefore refers not alone to the physical but, more commonly, to the inward eye. The expression, like the quality itself, has certainly always been more applicable to tactics, but it must also have its place in strategy. . . . The concept merely refers to the quick recognition of a truth that the mind would ordinarily miss or would perceive only after long study and reflection."
4. Lewis, *Mind and the World Order*, p. 390.
5. Jomini, *The Art of War*, p. 185.
6. Jomini, *The Art of War*, p. 306.

   For an interesting, if somewhat disjointed, discussion of the need for developing *coup d'oeil* in Air Force officers, see Captain Charles T. Barco, "Valuing Leadership in an Era of Prophets, Politicians, and Pugilists," *Airpower Journal*, Volume VIII, Number 3, Fall 1994, especially pp. 5-7 and 11-13.
7. Lewis, *Mind and the World Order*, p. 85.

7. Pushback

1. Paret, "Clausewitz," in *Makers of Modern Strategy*, p. 193.
2. Holley, *Technology and Military Doctrine*, p. 110.
3. Alger, *The Quest for Victory*, 174.
4. Warden, *The Air Campaign*, p. 116.
5. Jomini, *The Art of War*, p. 112.
6. Jomini, *The Art of War*, pp. 293-4.
7. Liddell Hart, *Strategy*, p. 5.
8. Quoted in Tsouras, *Warrior Words*, p. 63.

8. Is War Understandable?

1. *Joint Publication* 1-02, p. 113.

   The definition continued, "Although regular forces are often involved, irregular forces frequently predominate. Conflict often is protracted, confined to a restricted geographic area, and constrained in weaponry and level of violence. Within this state, military power in response to threats may be exercised in an indirect manner while supportive of other instruments of national power. Limited objectives may be achieved by the short, focused, and direct application of force."
2. Moseley, *A Philosophy of War*, p. 14.

   The concept of nonviolent conflict is easy enough to grasp; however, the concept of "non-violent hostilities" as opposed to "violent hostilities" is more difficult. From a practical standpoint, "hostilities" implies hostile actions, so the only nonviolent hostilities would seem to be hostile intentions.

3. Carl von Clausewitz, *On War*, Edited and Translated by Michael Howard and Peter Paret (Princeton, NJ: Princeton University Press, 1976), p. 579.
4. See Colin S. Gray, *Modern Strategy* (Oxford: Oxford University Press, 1999), especially pp. 55-68.

    See Clausewitz, *On War*, p. 605.
5. Van Creveld, *The Transformation of War*, p. 126.
6. Even with the qualifier "should be," this view will not, of course, be universally held.

    For instance, former US Senator Gary Hart would probably dissent. He wrote, "War is not an instrument of policy; it is a failure of policy. We cannot discuss the use of military power as an instrument of national policy without recognition that it is the lives of our sons and daughters that are most immediately at stake. . . . [Our] duty to our sons and daughters requires our policy makers to hold their lives in sacred trust." *The Fourth Power: A Grand Strategy for the United States in the Twenty-First Century* (Oxford: Oxford University Press, 2004), p. 162. One wonders what "use of military power" the Senator would find justifiable, i.e., that would not violate that "sacred trust."
7. Van Creveld, *The Transformation of War*, p. 166.
8. Smith, *The Utility of Force*, p. 5.
9. John Vasquez objected to the undefined term, "violence," in definitions of war, and argued that a definition of war must refer to deadly violence. "It would seem that war must involve organized violence that aims to kill members of another group, not simply to do them harm, otherwise war becomes too much like *force*." *The War Puzzle*, p. 23. (Emphasis in original.)
10. Hart, *The Fourth Power*, p. 108.
11. Sun Tzu, *The Art of War*, Translated by Samuel B. Griffith (London: Oxford University Press, 1963), p. 77. Although not as technically correct as the more recent Thomas Cleary translation, the Griffith translation carries more "martial ardor."

    Unless otherwise noted, all references to Sun Tzu's *The Art of War* will use the Griffith translation.
12. Sun Tzu, *The Art of War*, p. 71.
13. *Joint Publication 1*, p. I-1.

    The old DoD definition, "Fundamental principles by which the military forces or elements thereof guide their actions in support of national objectives," also noted that doctrine was "authoritative but requires judgment in application." *Joint Publication 1-02*, p. 168.
14. "Therefore, he who desires peace, let him prepare for war; he who wants victory, let him instruct his soldiers diligently; he who wishes favorable results, let him fight with skill, not with chance . . . ." Flavius Vegetius Renatus, *Epitoma Rei Militaris*, Preface to Book III, Edited with an English translation by Leo F. Stelton (New York: Peter Lang, 1990), p. 11.

    Alternate translations can also be instructive: "Therefore, he who desires peace, let him prepare for war. He who wants victory, let him train soldiers diligently. He who wants a successful outcome, let him fight with strategy, not at random." *Vegetius: Epitome of Military Science*, Preface to Book III, Translated with notes and introduction by N.P. Milner (Liverpool: Liverpool University Press, 1993), p. 62.

9. Has Anyone Understood War?

1. Sun Tzu, *The Art of War*, p. 84.
2. Edward C. Mann III, *Thunder and Lightning: Desert Storm and the Airpower Debates* (Maxwell AFB, AL: Air University Press, 1995), p. 17.
3. "Sun Tzu" means "Master Sun."

    The vowel sound of "Sun" is pronounced in the same way as the vowel sound of the

English word "soot," and the two words carry similar inflection. "Tzu" is pronounced "zuh," with a very sharp cutoff at the end.

"Sun Zi" is the modern Romanization.

Special thanks to Dr. David L. Rolston, Director of Publications at the University of Michigan Center for Chinese Studies, for help on this subject.

4. McNeilly, *Sun Tzu and the Art of Modern Warfare*, pp. 3-4.
5. See Keegan, *A History of Warfare*, especially pp. 202 and 221.
6. "War, however, is not the action of a living force upon a lifeless mass . . . but always the collision of two living forces." Clausewitz, *On War*, p. 77.
7. Van Creveld, *The Transformation of War*, p. 35.
8. Antoine Henri Jomini, *The Art of War*, Translated by G.H. Mendell and W.P. Craighill (Westport, CT: Greenwood Press, Publishers, undated), p. 115. Originally published in the US in 1862 by J.B. Lippincott & Co., Philadelphia, this volume contains the *Summary of the Art of War* as well as additional works.
9. "There can be little doubt that Kant had an influence on Clausewitz and *On War*, but there is no evidence that Clausewitz ever actually read any of Kant's works. Most historians believe that Clausewitz got his Kant second and third hand." Robert P. Pellegrini, *The Links Between Science, Philosophy and Military Theory: Understanding the Past, Implications for the Future* (Maxwell Air Force Base, AL: Air University Press, August 1997), p. 22.
10. See, for example, John I. Alger, *The Quest for Victory: The History of the Principles of War* (Westport, CT: Greenwood Press, 1982), especially p. 31. Originally published in 1890.
11. Alfred Thayer Mahan, *The Influence of Sea Power Upon History, 1660-1783* (Boston: Little, Brown, and Company, 1903 reprint), p. 2.
12. Alger, *The Quest for Victory*, pp. 88-92.

    See also Philip A. Crowl, "Alfred Thayer Mahan: The Naval Historian," in Peter Paret, Editor, *Makers of Modern Strategy: from Machiavelli to the Nuclear Age* (Princeton: Princeton University Press, 1986), p. 450.
13. Gray, *Modern Strategy*, pp. 230-1.
14. The OODA Loop contends that "one can depict all rational human behavior, individual and organizational, as a continual cycling through four distinct tasks: observation, orientation, decision, and action." *Air Force Doctrine Document 1-2, Air Force Glossary* (Maxwell AFB: Air Force Doctrine Center, 9 July 1999), p. 15. On-line, 20 July 2003, https://www.doctrine.af.mil/.
15. William S. Lind, "John Boyd's Art of War," *The American Conservative*, 16 August 2013. On-line, 26 February 2018, http://www.theamericanconservative.com/articles/john-boyds-art-of-war/.

10. Categorizing War

1. Keegan, *A History of Warfare*, especially chapter 1. See also Sun Tzu, *The Art of War*, p. 30.
2. See Moseley, *A Philosophy of War*, pp. 23ff, and Vasquez, *The War Puzzle*, pp. 51ff.
3. See 1 Samuel 17.
4. Francis Fukuyama points out, "Legitimacy is not justice or right in an absolute sense; it is a relative concept that exists in people's subjective perceptions. All regimes capable of effective action must be based on some principle of legitimacy." *The End of History and the Last Man* (New York: Perennial, 2002), p. 15. First published in 1993.
5. "The Islamic revival was . . . the nostalgic re-assertion of an older, purer set of values, said to have existed in the distant past, that were neither the discredited 'traditional values' of the recent past, nor the Western values that had been so poorly transplanted to the Middle East. In this respect, Islamic fundamentalism bears a more than superficial resemblance to European fascism." Fukuyama, *The End of History*, p. 236.

Earlier (p. 15), Fukuyama wrote, "Fascist ultranationalism maintained that the ultimate source of legitimacy was race or nation, specifically, the right of "master races" like the Germans to rule other people. . . . Nazism's assertion of German racial superiority had to be actively proven through conflict with other cultures." Substitute "religion" for "race," "Muslim" for "German," and "Al Qaeda" for "Nazism" and the statement seems equally valid.

6. Fukuyama, *The End of History*, pp. 235-6.
7. Of the 38,000 sorties, approximately 23,300 were strike missions directed against 7600 fixed and 3400 "flex" targets. *Kosovo/Operation Allied Force After-Action Report* (Washington, DC: US Department of Defense, 31 January 2000), pp. 79 and 87. Online, 11 October 2005, *www.defenselink.mil.*
8. Yugoslav forces captured three US soldiers near the Yugoslav-Macedonia border in the spring of 1999.
9. Daalder and O'Hanlon, "Unlearning the Lessons of Kosovo," pp. 128 and 131.

11. The Lure of Simplicity

1. ". . . numerous attempts were made to produce a harmony or a synthesis of their systems . . . . The characteristic elements of the two may be juxtaposed; they cannot be fused." David Knowles, *The Evolution of Medieval Thought*, Second Edition (London: Longman Group, 1988), p. 4.

   Indeed, "the attempt to reconcile opposites may lead to madness as well as to philosophy." Will and Ariel Durant, *The Story of Civilization, Vol. 10*, p. 486.

12. The Levels of War, as Theory

1. *Field Manual* 3-0, p. 2-2.
2. Van Creveld, *The Transformation of War*, p. 96. See also p. 206.
3. Clausewitz, *On War*, p. 128. (Emphasis in original.)

   See also Peter Paret, "Clausewitz," in Peter Paret, Editor, *Makers of Modern Strategy: from Machiavelli to the Nuclear Age* (Princeton: Princeton University Press, 1986), p. 190.
4. Jomini, *The Art of War*, pp. 11, 61-2, and 159.

   See also John Shy, "Jomini," in Peter Paret, Editor, *Makers of Modern Strategy: from Machiavelli to the Nuclear Age* (Princeton: Princeton University Press, 1986), pp. 167-8.
5. I owe thanks for this observation to LTC Tom Kratman, USA.
6. Liddell Hart, *Strategy*, p. 10.
7. Eccles, *Military Concepts and Philosophy*, p. 18.
8. Gray, *Modern Strategy*, p. 17. (Emphasis in original.)
9. See *DOD Dictionary of Military and Associated Terms*, pp. 219, 173, and 226, respectively.
10. Smith, *The Utility of Force*, p. 14.
11. *Joint Publication 1*, 14 November 2000, p. III-1. Unfortunately, the term was left undefined and appears never to have been incorporated into the official list of Joint terminology.
12. Eccles, *Military Concepts and Philosophy*, p. 18.

    See also Steven Metz, "The Operational Level of Nuclear War Fighting: Missing or Unnecessary?", *Airpower Journal*, Volume VII, Number 1, Spring 1993, p. 13; and Mann, *Thunder and Lightning*, p. 168.
13. Miyamoto Musashi, *Go Rin No Sho* (*A Book of Five Rings*), Translated by Victor Harris (Woodstock, NY: Overlook Press, 1974), p. 48.

    He continues, "To become over-familiar with one weapon is as much a fault as not knowing it sufficiently well. You should not copy others, but use weapons which you can handle properly. It is bad for commanders and troopers to have likes and dislikes."

14. See, for example, Phillip S. Meilinger, *Ten Propositions Regarding Air Power* (Air Force History and Museums Program, 1995), p. 8.
15. Sun Tzu, *The Art of War*, p. 77.
16. Compare p. 40 with p. 18 of Meilinger, *Ten Propositions Regarding Air Power*.
17. Many thanks to Lieutenant Colonel Erik Eliasen for discussions of this topic.
18. *Air Force Doctrine Volume 1, Basic Doctrine*, 27 February 2015. (No page numbers in online document.)
19. *Joint Publication* 3-0, pp. I-13 - I-14.
20. *Joint Publication* 3-0, p. V-5. It is interesting that previous guidance aimed a campaign at achieving a single strategic or operational objective, rather than both.
21. From p. V-1 of the now-superseded 2000 version of *Joint Publication 1*, which continued: "The [Joint Force Commander] uses [the campaign] to arrange tactical, operational, and strategic actions into a series of related major operations, focused on accomplishing strategic and operational objectives." This was a curious construction. It seems more reasonable that the commander would arrange the actions into the campaign; i.e., that the campaign would *be* the "series of related major operations."
    These distinctions were dropped from the 2007 version of JP-1.
22. *MCDP 1, Warfighting*, pp. 15-7.
    With thanks to LTC Tom Kratman, USA, for this observation, it is interesting to note that for a time the Marine Corps included the OODA Loop in its thinking, but Boyd's cycle has more recently fallen out of favor. It remains to be seen if his physical, mental, and moral "levels" of war will also fade from USMC usage.
23. William S. Lind and LtCol Gregory A. Thiele, USMC, *4th General Warfare Handbook* (Kouvola, Finland: Castalia House, 2015), p. 13.
24. Lind, "John Boyd's Art of War."

13. The Media of War, as Theory

1. Daalder and O'Hanlon, "Unlearning the Lessons of Kosovo," p. 128.
2. Quoted in Major Mark K. Dippold, USAF, "Air Occupation: Asking the Right Questions," *Airpower Journal*, Volume XI, Number 4, Winter 1997, p. 73. Dippold suggests the term "air intervention" and points out that air occupation may have different objectives than a ground occupation.

14. More Media of War: The Possibility of Space

1. The "Space Commission" was also known by its full name, the "Commission to Assess United States National Security Space Management and Organization," and by the name of its chairman: the Rumsfeld Commission. *Report of the Commission to Assess United States National Security Space Management and Organization* (Washington, DC: January 11, 2001). On-line, 3 July 2011, *http://www.dod.gov/pubs/space20010111.html*.
2. Major Shawn P. Rife, "On Space-Power Separatism," *Airpower Journal*, Volume XIII, Number 1, Spring 1999, p. 21.
    The Western Development Division, precursor to today's Space and Missile Systems Center at Los Angeles Air Force Base, was founded in 1954.
3. Colin S. Gray and John B. Sheldon, "Space Power and the Revolution in Military Affairs—A Glass Half Full?", *Airpower Journal*, Volume XIII, Number 3, Fall 1999, p. 27.
    Some of Keegan's analysis of the development of navies and the proximity of naval

battles to ports and sea lanes can be extrapolated to the nascent conquest of space. Keegan, *A History of Warfare*, pp. 64-8.

See also Lieutenant Colonel Cynthia A.S. McKinley, "The Guardians of Space—Organizing America's Space Assets for the Twenty-First Century," *Aerospace Power Journal*, Volume XIV, Number 1, Spring 2000, pp. 37-45.

4. See Liddell Hart, *Strategy*, especially p. 125, for discussion on railroads and their effects on armies.
5. B.H. Liddell Hart, *History of the Second World War* (New York: Perigee Books, 1982), p. 158. (Original copyright 1970.)
6. "Space Policy," *Department of Defense Directive* 3100.10, 9 July 1999, p. 2.

   See also "Space Policy," *Department of Defense Directive* 3100.10, 18 October 2012, incorporating Change 1, 4 November 2016. On-line, 30 April 2018, *https://www.hsdl.org/?abstract&did=796643*.
7. Richard Szafranski and Martin C. Libicki, "'. . .Or Go Down in Flame?' Toward an Airpower Manifesto for the Twenty-First Century," *Airpower Journal*, Volume X, Number 3, Fall 1996, p. 71.
8. Gray and Sheldon, "Space Power and the Revolution in Military Affairs," p. 25.

   The electromagnetic spectrum as a medium of conflict will not be dealt with here, since electromagnetic energy traverses other media but is not a physical realm unto itself; however, the realm of "information" may require adjustments to the use of strict physical definitions. See the discussion of cyberspace and information warfare in the next chapter.
9. The Air Force moved from *Basic Aerospace Doctrine* (1992), which contained a chapter on the aerospace concept, to *Air Force Basic Doctrine* (1997), which used the term "air and space" exclusively. Major Carl Baner, "Defining Air and Space Power," *Aerospace Power Chronicles*, 11 March 1999. On-line, 12 April 2004, *http://www.airpower.maxwell.af.mil*.

   Just two years later, to mark the turn of the millennial year, the Air Force changed its professional journal from *Airpower Journal* to *Aerospace Power Journal*; then it changed the name once again, this time to *Air and Space Power Journal*.

   See also Meilinger, *Ten Propositions Regarding Air Power*, pp. 13, 57.

   See also Mowbray, "Air Force Doctrine Problems: 1926-Present," pp. 21-41.
10. *Air Force Doctrine Volume 1, Basic Doctrine* (Maxwell Air Force Base, AL: United States Air Force, 27 February 2015). On-line, 5 March 2018, http://www.doctrine.af.mil/Core-Doctrine/Vol-2-Leadership/. No page numbers in online document.
11. Sun Tzu, *The Art of War*, p. 117, et al.
12. The "grammar of war" is a Clausewitzian idea. See Gray, *Modern Strategy*, pp. 206-27 (for the "grammar of war" of the land and sea) and 228-72 (for the "grammar of war" of air, space, and cyberspace).
13. As noted in Chapter Two, this quote is from Dr. W. Edwards Deming.

    For more discussion on developing "aerospace" doctrine, see Mowbray, "Air Force Doctrine Problems: 1926-Present."

    See also note 7, above.
14. Builder, *The Icarus Syndrome*, p. 284.
15. ". . . though there is an indistinct interface between air and space between 50 and 75 nautical miles, there is a distinct difference between the physics of aviation and spaceflight. This interface is similar to the land-sea interface. The location of the land-sea interface is more distinct than the air-space interface, but the physics of land operations are significantly different from sea operations." Michael R. Mantz, *The New Sword: A Theory of Space Combat Power* (Maxwell AFB, AL: Air University Press, 1995), p. 34.

    The argument that a future aerospace plane will make the "aerospace continuum" a reality does not account sufficiently for the fact that such a vehicle will be a hybrid: *capable* of operating in two different environments, but optimized for neither. See Rife, "On Space-Power Separatism," p. 25.

I am indebted to Captain Mike Scardera for discussion about this point.

16. Electronic signals are affected by gravity, but only to a miniscule extent. Depending on the wavelength, they may be affected much more by weather (especially rain).
17. *Air Force Doctrine Document 1*, 17 November 2003, p. 4.
18. *Air Force Doctrine Document 1*, 14 October 2011, p. 12.
19. *Air Force Doctrine Document 1*, 14 October 2011, p. 13. Verbatim in the 2015 *Basic Doctrine* also.
20. Taylor Dinerman, "'Near Space:' a new area of operations or a new Pentagon buzzword?", *The Space Review*, 20 September 2004. On-line, 13 April 2005, *http://www.thespacereview.com/article/230/1*.

    See also "Near-space demo aims to prove boost in combat communication ability," Air Force Space Command Press Release (Colorado Springs, CO: Air Force Space Command News Service, 8 March 2005). On-line, 11 May 2005, *http://www.peterson.af.mil/hqafspc/News/News_Asp/nws_tmp.asp?storyid=05-070*.

    The term "high altitude long endurance" is from Jeffery R. Barnett, *Future War: An Assessment of Aerospace Campaigns in 2010* (Maxwell Air Force Base, AL: Air University Press, January 1996), p. 46. On-line, 18 November 2004, *http://www.maxwell.af.mil/au/aul/aupress/*.
21. "Near-space demo aims to prove boost in combat communication ability."

15. Still More Media of War: The Possibility of Cyberspace

1. *Joint Publication 1*, 14 May 2007, p. IV-15.
2. Sun Tzu, *The Art of War*, p. 112. This guidance, which applies when the commander at the head of the troops is assumed to have a better grasp of the immediate situation than the sovereign, is most famously illustrated in the (probably apocryphal) story of the drill and ceremonies Master Sun demonstrated with the Emperor's concubines.
3. *DOD Dictionary of Military and Associated Terms* (Washington, DC: US Department of Defense, February 2018), pp. 60 and 111. On-line, 26 February 2018, http://www.jcs.mil/Doctrine/.

16. Combining the Levels and Media of War

1. See "Three Levels of War," in *Air Force Manual 1-1, Volume II*, pp. 43-9.
2. For a thorough discussion of this, see Mantz, *The New Sword*.

    See also Lieutenant General Bruce Carlson, "Protecting Global Utilities: Safeguarding the Next Millennium's Space-Based Public Services," *Aerospace Power Journal*, Volume XIV, Number 2, Summer 2000, pp. 37-41; and Ralph Millsap and D.B. Posey, "Organizational Options for the Future Aerospace Force," *Aerospace Power Journal*, Volume XIV, Number 2, Summer 2000, pp. 48-52.
3. Billy M. Knowles, "Hold, Withdraw, or Advance: The Role of Tactical Airlift in the Equation," *Airpower Journal*, Volume III, Number 2, Spring 1989, p. 47. (Emphasis in original.)
4. See Mantz, *The New Sword*, especially pp. 31-35.
5. Major General (select) John L. Barry and Colonel Darrell L. Herriges, "Aerospace Integration, Not Separation," *Aerospace Power Journal*, Volume XIV, Number 2, Summer 2000, pp. 42-47 (especially p. 42).
6. The description of levels of war being associated with particular weapons is from Eccles, *Military Concepts and Philosophy*, p. 18.
7. As indicated, this is vastly different from the accepted division of responsibilities. See "Three Levels of War," in *Air Force Manual 1-1, Volume II*, p. 43.
8. See Steven Metz, "The Operational Level of Nuclear War Fighting: Missing or Unnecessary?", pp. 13-20.

9. Hart, *The Fourth Power*, p. 107.
10. Liddell Hart, *Strategy*, p. 322.
    General Sir Rupert Smith refers to the grand strategic level as the "political" level of war. *The Utility of Force*, p. 10.
11. Liddell Hart, *Strategy*, pp. 322 and 353.
12. Gray Rinehart (editor), Erik Eliasen, Steve Julian, Mike Lutton, Keith Phillips, John Taylor, Mark VanVoorhis, and James H. Galt-Brown, "Toward Space War," *High Frontier*, Volume 1, Number 3, Winter 2005, p. 47.
13. Notably Senator Gary Hart, in his thought-provoking book *The Fourth Power*.
    Senator Hart's book is essentially a lengthy essay, and presents some interesting and worthy observations such as, "We are not just a democracy of rights; we are principally a republic of duties. Further, our nature as a republic presupposes a common good, a commonwealth, a definable national interest greater than a mere collection of narrow or special interests" (p. 152).
    Unfortunately, the book does not always achieve the level of statesmanship the Senator attempted, falling often into criticism and positing a "secret dream of empire" hidden in the stated objectives of the Terror War (p. 99). And, in an odd but presumably honest mistake, the book misquotes the Constitution on p. 38: "a central purpose of the [US] Constitution itself was to 'ensure justice.'" The Preamble actually states that the Constitution was "ordain[ed] and establish[ed]" to "establish justice," not to ensure it.
14. See *MCDP 1, Warfighting*, and *4th Generation Warfare Handbook*.
15. Senator Gary Hart suggests that "A new grand strategy for the United States in the twenty-first century should assume the following characteristics: it will be multidimensional; it will transcend impulse, reaction, and ideology; it will rely on appropriate means to achieve justifiable ends; it will recognize its own constraints; and it will be transparent." Further, it must have three other traits: consistency, coherence, and adaptability. *The Fourth Power*, p. 14.
16. *Joint Publication 1*, p. I-5.

17. Quasi-Theories

1. John A. Warden, *The Air Campaign: Planning for Combat* (Washington: Pergamon-Brassey's, 1989), p. 142.
2. Paret, "Clausewitz," in *Makers of Modern Strategy*, p. 199.
3. Clausewitz, *On War*, p. 101.
    In fact, van Creveld refers to fog *as* uncertainty in *The Transformation of War*. See pp. 104ff.
4. Observation made by Colonel William Perkins, Air Force Space Command, in mid-1994.
5. Paret, "Clausewitz," in *Makers of Modern Strategy*, p. 202.
6. Clausewitz, *On War*, p. 119.
7. Colonel John A. Warden III, "The Enemy as a System." *Airpower Journal*, Volume IX, Number 1, Spring 1995, p. 42.
8. Keegan, *A History of Warfare*, p. 63.
9. See, e.g., van Creveld, *The Transformation of War*, p. 31.
10. Unfortunately the author did not get the name of the gentleman, who was another patient on a Medivac flight that had an extended RON (rendezvous overnight) at Scott AFB, Illinois, in January 1997.
    See also Colonel Owen Jensen, "Information Warfare: Principles of Third-Wave War," *Airpower Journal*, Volume VIII, Number 4, Winter 1994, p. 39: "A military most vulnerable to information warfare strategy is from an information society. . . . The more we depend on our sensors and computers and space-based communications, the easier we are to unplug."

See also R.L. DiNardo and Daniel J. Hughes, "Some Cautionary Thoughts on Information Warfare," *Airpower Journal*, Volume IX, Number 4, Winter 1995, p. 75; and Major Richard W. Aldrich, "The International Legal Implications of Information Warfare," *Airpower Journal*, Volume X, Number 3, Fall 1996, p. 100.

Special thanks to Captain Mike Scardera for contributing to this section.

11. See Warden, "The Enemy As A System," especially pp. 44-8. Colonel Warden applies the model to strategic warfare on pp. 49-54 of the article.

    Thanks to Colonel Michael Mantz for pointing out how the five rings theory hearkens back to the nuclear warfighting theory of decapitation. This is also mentioned is Richard Szafranski, "The Problems With Bees and Bombs," *Airpower Journal*, Volume IX, Number 4, Winter 1995, p. 94; and Mann, *Thunder and Lightning*, pp. 27-8 and 44.

    See also Major Russell J. Smith, "Developing an Air Campaign Strategy," *Aerospace Power Chronicles*, 23 November 1999. On-line, 12 April 2004, *http://www.airpower.maxwell.af.mil*.

12. For more on nodal analysis, see Major Steven M. Rinaldi, *Beyond the Industrial Web: Economic Synergies and Targeting Methodologies* (Maxwell AFB, AL: Air University Press, April 1995), especially Chapter Three. On-line, 13 October 2005, *http://www.maxwell.af.mil/au/aul/aupress/*.
13. See Liddell Hart, *Strategy*, especially p. 333.
14. Lewis Ware, "Some Observations of the Enemy As a System," *Airpower Journal*, Volume IX, Number 4, Winter 1995, p. 91.

    The system flowchart describes processes and plots interactions. For basics on the system flowchart, see the author's "A New Paradigm for Organizational Structure." *Airpower Journal*, Volume VI, Number 1, Spring 1992, pp. 43-53.

15. Warden, "The Enemy As a System," p. 43.
16. Jensen, "Information Warfare: Principles of Third-Wave War," p. 39.
17. Jomini, *The Art of War*, pp. 35-6 and 250.
18. *The Commission on the Intelligence Capabilities of the United States Regarding Weapons of Mass Destruction*,

    31 March 2005, p. 125. (Emphasis in original.) On-line, 30 April 2005; *http://www.wmd.gov/report/*.

19. Dr. W. Edwards Deming. The follow-on question is, "Does that world make contact with ours?"
20. Van Creveld, *The Transformation of War*, p. 173.
21. "Schrodinger's Cat" refers to a thought experiment devised by physicist Erwin Schrodinger. In the experiment, the condition (i.e., living or dead) of a cat is indeterminate without direct observation. The point of the experiment is that, at a quantum level, observations affect the things observed.

    The experiment follows this pattern: a cat is placed inside a box with a cylinder of poison gas, a small amount of radioactive material, and a trigger that will release the poison gas if the radioactive material decays. The question is then posed, is the cat alive or dead? The answer is, *we do not know until we open the box*. The state of the cat is indeterminate so far as we are concerned—it is, in a sense, both alive and dead—and the cat's "actual" state is determined as much by our observation as by the prevailing conditions.

22. Alger, *The Quest for Victory*, p. 3.
23. As an example, Captain Mike Scardera pointed out that Stalin consolidated power by genocide. However, what is considered nefarious in one age may end up being de rigueur in another; e.g., submarine warfare.
24. "The most basic weakness [of the Soviet Union] whose full gravity escaped the attention of Western observers was economic. It was much more difficult to tolerate economic failure in the Soviet system because the regime itself had explicitly based its claims to legitimacy on its ability to deliver its people a high material standard of living. . . . Flat overall GNP growth, coupled with yearly increases in defense spending

through the early 1980s of 2 to 3 percent, meant that the civilian economy was actually shrinking at an appreciable rate for the decade before Gorbachev came to power." Fukuyama, *The End of History*, pp. 28-9.

25. See, e.g., Evan Thomas, Rod Nordland, and Christian Caryl, "Operation Hearts and Minds," *Newsweek*, Volume 143, Number 1, 29 December 2003-5 January 2004, pp. 34-43.

18. Considering the "Principles of War"

1. *Joint Publication 1*, 2007 version, p. I-2.

    Interestingly, where the "capstone" doctrine in 2007 referred to "experience across the range of military operations," current subordinate doctrine—specifically, operations doctrine—credits "experience gained in a variety of [irregular warfare] situations" for reinforcing the value of the three added principles (restraint, perseverance, and legitimacy). *Joint Publication 3-0*, 2017, p. I-2.
2. *ADP 1-01, Doctrine Primer*, p. 4-7.
3. *Field Manual* 3-0, 2008, pp. 5-4 and A-1. Compare *Field Manual 3-0, Operations* (*Change 1*) (Washington, DC: Department of the Army, 6 December 2017). On-line, 10 May 2018, https://armypubs.army.mil/ProductMaps/PubForm/FM.aspx.

    2001 version of FM 3-0 stated that "The principles are the enduring bedrock of Army doctrine," that they "guide and instruct commanders as they combine the elements of combat power," and that they "reflect the distillation of Army experience into a set of time-tested guidelines," none of which statements appear in the 2008 version.
4. In the 2008 FM 3-0, the principles were relegated to an appendix, while the elements of combat power were treated in their own chapter. The 2001 manual listed the principles after the elements of combat power.
5. *ADP 1-01*, p. 4-7.
6. "Principles of War," in *Air Force Manual 1-1, Volume II*, p. 14.

    See also *Air Force Doctrine Document 1*, September 1997, which contained a much more focused discussion than its predecessor, treating each principle from the "airman's perspective," a trend continued in the 2003 version.
7. *Air Force Doctrine Volume 1*, 27 February 2018. (No page numbers in on-line version; however, the same text was on p. 30 of the 2011 version.)
8. *MCDP 1-0*, p. A-2.
9. Jomini, *The Art of War*, p. 304.
10. Clausewitz, *On War*, p. 117.
11. Clausewitz, *On War*, p. 134.
12. Russell F. Weigley, *The American Way of War: A History of United States Military Strategy and Policy* (New York: Macmillan Publishing Company, Inc., 1973), p. 215.
13. For example, Colonel Warden's repeated invocation of mass, concentration, and economy of force. Warden, *The Air Campaign*, pp. xiii, xx, 29, 99, and 143.
14. See Alger, *The Quest for Victory*, especially pp. 195-270.

    For Jomini, "Principles were guides to action, not infallible mathematical calculations. The specific application of principles would vary," because he had recognized in his early work (1807) that "in warfare the unexpected must be expected . . . ." Shy, "Jomini," in *Makers of Modern Strategy*, pp. 154, 167.
15. Alger, *The Quest for Victory*, p. xxii.
16. Around the turn of the century, *Joint Publication 3-07, Joint Doctrine for Military Operations Other Than War* (Washington, DC: US Department of Defense, 16 June 1995) described MOOTW and catalogued its principles. Found on-line, 28 February 2002, at *http://www.dtic.mil/doctrine/s_index.html*, the document is no longer part of the Joint canon.

17. Robert S. Frost, *The Growing Imperative to Adopt "Flexibility" as an American Principle of War* (Carlisle, PA: Army War College Strategic Studies Institute, 15 October 1999). See especially pp. 36 (note 6) and 46 (note 65).

    On the importance of logistics, see Richard Holmes, "Battle: The Experience of Modern Combat," in Charles Townshend, editor, *The Oxford History of Modern War* (Oxford: Oxford University Press, 2000), especially p. 236.

    The three additional "principles of joint operations" are found in JP 3-0, pp. A-3 - A-5, and FM 3-0, pp. A-3 - A-4.
18. Warden, *The Air Campaign*, p. 143.
19. Warden, *The Air Campaign*, p. 99.
20. Mahan, *The Influence of Sea Power Upon History*, p. 7.
21. Thomas E. Griess, "A Perspective on Military History," in *A Guide to the Study and Use of Military History* (US Army Center of Military History, 1982 reprint), p. 34.

    Colonel Griess continues to defend the principles, however: "If an historically based principle is fallible, however, it is infinitely better than pure theory ungrounded on historical experience."
22. Holley, *Technology and Military Doctrine*, pp. 110 and 119.

    Despite its interesting structure, which interweaved Master Sun's text and commentaries on the text with ancient and contemporary battle examples, this seems to be the chief drawback of McNeilly's *Sun Tzu and the Art of Modern Warfare*: carefully selected examples that bolster the argument, without apparent regard to possible counterexamples.
23. Holley, *Technology and Military Doctrine*, p. 96.
24. Mahan, *The Influence of Sea power Upon History*, p. 9.
25. Doing so runs the risk of doing "dreadful injury to the facts." John Keegan, *The Face of Battle* (New York: Viking Press, 1976), p. 22.
26. *Air Force Doctrine Volume 1*, 27 February 2018. (No page numbers in on-line version; however, the same text was on p. 30 of the 2011 version.)

19. The Nine Classical Principles of War

1. *Joint Publication* 3-0, p. A-1. All the quotations in this paragraph are from the same source.
2. Quoted in Tsouras, Editor, *Warriors' Words*, p. 34.
3. *Joint Publication* 3-0, p. A-1.
4. Hart, *Strategy*, p. 329.
5. Since any prediction is based on theory, and all knowledge is probable only, as pointed out in Chapters 4 and 5.
6. *Joint Publication* 3-0, p. A-1.

    The old FM-1 stated that the Army's operational concept was to "seize, retain, and exploit the initiative" through "speed, shock, surprise, depth, simultaneity, and endurance." (*Field Manual 1*, p. 3-4.) Interestingly, that description appears to have fallen out of use.
7. Sun Tzu, *The Art of War*, p. 85.
8. Robert L. O'Connell, *Of Arms and Men: A History of War, Weapons, and Aggression* (Oxford: Oxford University Press, 1989), p. 233.
9. John Keegan, *The Face of Battle*, p. 230. (Emphasis in original.)
10. One who did was former Navy Secretary John Lehman, who wrote, "President Clinton described the attack on Cole as an act of cowardice and of terrorism. It was of course neither. It was a well-planned act of war by obviously brave and disciplined warriors. . . ." John Lehman, "USS Cole: An Act of War," *Washington Post*, 15 October 2000, p. B-7. On-line, 5 October 2005, *http://www.washingtonpost.com*.
11. *Joint Publication* 3-0, p. A-2.
12. *Joint Publication* 3-0, p. A-2.

13. *DOD Dictionary of Military and Associated Terms*, p. 42.
14. *Joint Publication* 3-0, p. A-2.
15. For example, Mahan stressed concentration in his adaptation of Jomini's principles. Weigley, *The American Way of War*, p. 174.
16. Jomini, *The Art of War*, p. 63.

    Later, in a supplement to his *Summary of the Art of War*, he expanded this "one great principle" into "two fundamental principles: the first being, *to obtain by free and rapid movements the advantage of bringing the mass of the troops against fractions of the enemy*; the second, *to strike in the most decisive direction* . . . . The whole science of great military combination is comprised in these two fundamental truths." *The Art of War*, p. 299. (Emphasis in original.)
17. Mahan, *The Influence of Sea Power Upon History*, p. 381.
18. "Mahan, suspicious as always of new technology, advocated an investment in more ships of smaller size [rather than only a few Dreadnoughts]. He debated the issue in the pages of the *Naval Institute Proceedings* . . . ." Crowl, "Alfred Thayer Mahan: The Naval Historian," in *Makers of Modern Strategy*, p. 472.

    See also O'Connell, *Of Arms and Men*, p. 228.
19. See O'Connell, *Of Arms and Men*, especially chapter 15.
20. See especially Warden, *The Air Campaign*.

    See also Shy, "Jomini," in *Makers of Modern Strategy*, p. 182.
21. Thanks to Colonel Michael Mantz for bringing up this point.

    See also Lieutenant Colonel Edward Mann, "One Target, One Bomb: Is the Principle of Mass Dead?" *Airpower Journal*, Volume VII, Number 1, Spring 1993, pp. 35-43.
22. Colonel Meilinger's 7th Proposition. Meilinger, *Ten Propositions Regarding Air Power*, pp. 41-8.
23. Sun Tzu, *The Art of War*, p. 122.

    This despite, e.g., Master Sun's advice on p. 79 that one may attack if having five times the numbers of one's opponents.
24. *Vegetius: Epitome of Military Science*, Book I, p. 2.

    The Stelton translation puts it this way, "For in the contest of battle a few trained men are more ready for victory, whereas an untried and unskilled multitude is always subject to slaughter." Vegetius, *Epitoma Rei Militaris*, Book I, p. 11.
25. Clausewitz, *On War*, pp. 134-5. (Emphasis in original.)
26. Jomini, *The Art of War*, p. 151.
27. See Shy, "Jomini," in *Makers of Modern Strategy*, p. 169.
28. O'Connell, *Of Arms and Men*, p. 78.

    See also Liddell Hart, *Strategy*, p. 28.
29. See Gordon A. Craig, "Delbruck: The Military Historian," in Peter Paret, Editor, *Makers of Modern Strategy: from Machiavelli to the Nuclear Age* (Princeton: Princeton University Press, 1986), p. 326.
30. Jared Diamond, *Guns, Germs, and Steel* (New York: W.W. Norton and Company, 1998), pp. 67-81, and especially pp. 74ff.
31. Will and Ariel Durant, *The Story of Civilization, Vol. 10: Rousseau and Revolution* (New York: MJF Books, 1967), p. 458.
32. Walter Pintner, "Russian Military Thought: The Western Model and the Shadow of Suvorov," in Peter Paret, Editor, *Makers of Modern Strategy: from Machiavelli to the Nuclear Age* (Princeton: Princeton University Press, 1986), p. 360.

    Professor Pintner points out this was probably because the need to use portions of those troops to defend other areas "dissipated Russian numerical strength." This implies that if the Russians had thrown all their forces into the melee, they could not have lost, an implication based on faith in the power of mass.
33. Liddell Hart, *Strategy*. See especially pp. 48-9, 91, 253, 260, and 315.
34. Jomini, *The Art of War*, p. 295.
35. Jomini, *The Art of War*, p. 300.

36. Liddell Hart, *Strategy*, p. 328.
37. *Joint Publication* 3-0, p. A-2.
    This principle gained special popularity as the military climate grew more austere. See, as an example, Lieutenant Colonel Mark A. Gunzinger, "Airpower as a Second Front," *Airpower Journal*, Volume IX, Number 3, Fall 1995, pp. 63-73.
38. Mann, *Thunder and Lightning*, p. 64.
    This idea was also presented by Sun Tzu. See *The Art of War*, p. 98.
39. *Joint Publication 1*, 14 November 2000, p. III-7.
    That document was superseded by the 14 May 2007 edition, and the quoted statement does not appear in that or the newer 2013/2017 edition.
40. Sun Tzu, *The Art of War*, p. 98.
41. Proposed by the author in an unpublished monograph.
42. Musashi, *Go Rin No Sho*, p. 45.
43. *Joint Publication* 3-0, p. A-2.
44. Jomini, *The Art of War*, p. 304.
45. John Keegan, *The Second World War* (New York: Viking Penguin, 1989), pp. 58-9.
    See also Liddell Hart, *Strategy*, p. 161.
46. Special thanks to Doctor James H. Galt-Brown for this point.
47. Sun Tzu, *The Art of War*, p. 102.
48. Alger, *The Quest for Victory*, p. 114.
49. Sun Tzu, *The Art of War*, p. 134.
    Also: "While we have heard of blundering swiftness in war, we have not yet seen a clever operation that was prolonged." *The Art of War*, p. 73.
50. Meilinger, *Ten Propositions Regarding Airpower*, pp. 31-2.
51. Sun Tzu, *The Art of War*, p. 73.
    In addition, Chen Hao wrote: "As it is said, be swift as the thunder that peals before you have a chance to cover your ears, fast as the lightning that flashes before you can blink your eyes." Sun Tzu, *The Art of War*, Translated by Thomas Cleary (Boston: Shambhala, 1988), p. 58.
52. See, for example, Robert L. Forward, "Surgical Strikes from Space Using Solar Sail Statites," in Robert L. Forward, *21st Century Space Propulsion Study Addendum* (Phillips Laboratory Propulsion Directorate Technical Report PL-TR-91-3022, June 1991).
53. McNeilly, *Sun Tzu and the Art of Modern Warfare*, p. 96.
54. *Field Manual 1*, p. 3-4.
55. *Joint Publication* 3-0, p. A-2.
56. Thanks to Colonel William Perkins for this observation.
57. H. Norman Schwarzkopf, with Peter Petre, *It Doesn't Take a Hero* (New York: Linda Grey Bantam Books, 1992), p. 313.
58. See *Joint Publication 1*, especially chapter 4.
59. Thanks (again) to Captain Mike Scardera for contributing to this section.
60. *Joint Publication* 3-0, p. A-3.
61. John Keegan points out that the Poles first broke the Germans' Enigma codes, and purely by mathematical reasoning. *Intelligence in War: Knowledge of the Enemy from Napoleon to Al-Qaeda* (New York: Alfred A. Knopf, 2003), pp. 153-4.
62. Keegan, *Intelligence in War*, pp. 167ff.
63. See Clausewitz, *On War*, p. 84, and Jomini, *The Art of War*, p. 250.
64. Sun Tzu, *The Art of War*, p. 145.
65. *The Commission on the Intelligence Capabilities of the United States . . .*, p. 164.
66. Sun Tzu, *The Art of War*, p. 66.
67. Liddell Hart, *Strategy*, p. 49.
68. Liddell Hart, *History of the Second World War*, pp. 68 and 119.
69. George C. Kenney, *General Kenney Reports: A Personal History of the Pacific War* (Washington, DC: Reprinted by the Office of Air Force History, 1987), pp. 262, 267, and 501-2.

70. *Joint Publication* 3-0, p. A-3.
71. *Field Manual* 3-0, p. 2-41.
72. See Judges 7.
73. My appreciation to Colonel William Perkins for the emphasis on our vulnerabilities.
74. *Joint Publication* 3-0, p. A-3.
75. See John Keegan, *The Mask of Command* (New York: Penguin Books, 1987), pp. 98, 100, and 151.
76. Jomini, *The Art of War*, p. 179.

20. Beyond the Classical Principles of War

1. *Joint Publication* 3-07, p. viii.

   The first three MOOTW principles included in this now-superseded publication were presumably not carried over to the new Joint Pub because they were too close in character to the traditional principles. The first, "Direct every military operation towards a clearly defined, decisive and attainable objective," was—like its counterpart among the "traditional" principles—less a principle than a necessity for military operation of any type. The second seemed analogous to unity of command ("Unity of effort in every operation ensures all means are directed to a common purpose") and the third mirrored a traditional counterpart as well ("Security is always important and depends on never permitting hostile factions to acquire a military, political, or informational advantage").
2. *Joint Publication* 3-0, p. A-3.
3. Specifically, the 2010 version.
4. *Joint Publication* 3-0, p. A-4.
5. *Joint Publication* 3-0, p. A-4.
6. Sun Tzu, *The Art of War*, p. 73.
7. *Joint Publication* 3-0, p. A-4.
8. *Joint Publication* 3-0, p. A-4.
9. *Joint Publication* 3-0, p. A-4.
10. As stated in the 2010 version of *Joint Publication* 3-0, p. A-4.
11. *Air Force Doctrine Volume 1, Basic Doctrine* (Maxwell Air Force Base, AL: United States Air Force, 27 February 2015). On-line, 5 March 2018, http://www.doctrine.af.mil/Core-Doctrine/Vol-1-Basic-Doctrine/. (No page numbers in online document.)
12. Tom Kratman, "Indirectly Mistaken Decision Cycles," Baen.com, 2012. On-line, 20 June 2018, *https://www.baen.com/decisioncycles*. All quotes to the end of the chapter are from this source.

21. The Possibility of Error

1. Suggested by Colonel Michael Mantz.
2. The Johari Window was originally developed by psychologists to illustrate the dynamics of human relationship. It consisted of a 2x2 matrix of "Known to Others, Known to Self", "Known to Others, Unknown to Self", "Unknown to Others, Known to Self", and "Unknown to Others, Unknown to Self."
3. *Air Force Doctrine Volume* 1, 27 February 2015. (No page numbers in online document.)
4. Jomini, *The Art of War*, p. 112.
5. Jomini, *The Art of War*, p. 64.
6. Jomini, *The Art of War*, p. 112. Jomini attributed exceptions "to faults of execution, entirely foreign to the principles in question."
7. Clausewitz, *On War*, p. 593.

8. John F. Votaw, "An Approach to the Study of Military History," in *A Guide to the Study and Use of Military History* (US Army Center of Military History, 1982 reprint), p. 49.

   Thanks to Second Lieutenant Ann Brosier, USMA '95, for confirmation and elaboration.
9. Alger, *The Quest for Victory*, pp. 128-9, 141.
10. Jomini, *The Art of War*, p. 296.

22. Complexities and Interactions

1. Clausewitz, *On War*, p. 157.
2. Clausewitz, *On War*, p. 158.
3. *Air Force Doctrine Volume 1*, 27 February 2015. (No page numbers in online document.)
4. "Principles of War," in *Air Force Manual 1-1, Volume II*, p. 9.
5. *Joint Publication 1*, 14 November 2000, p. III-7.
6. Special thanks to Dr. John Surak, Clemson University, for emphasizing the importance of interactions in a number of applications.

   The interactions between the media of war (as opposed to their combination, which will be presented in the next chapter) can also be mapped out and studied; special thanks to Captain Mike Scardera for challenging discussions of this theory.
7. Donald J. Wheeler, *Understanding Industrial Experimentation*, 2nd Edition (Knoxville, TN: SPC Press, 1988), p. 135.
8. See Wheeler, *Understanding Industrial Experimentation*; and Ronald D. Moen, Thomas W. Nolan, and Lloyd P. Provost, *Improving Quality Through Planned Experimentation* (New York: McGraw-Hill, 1991).
9. See Alger, *The Quest for Victory*.
10. These three "principles" were included in Air Force doctrine as late as 1984, but did not appear in doctrine published after 1992. "Principles of War," in *Air Force Manual 1-1, Volume II*, p. 14.
11. *Air Force Doctrine Volume 1*, 27 February 2015. (No page numbers in online document.)
12. Clausewitz, *On War*, p. 595.

23. From Analysis to Synthesis

1. Perhaps instead of a plus sign, a union symbol would have been a better choice; although a case may be made that an intersection would have been better.
2. Columbia *Accident Investigation Board Report Volume 1* (Washington, DC: National Aeronautics and Space Administration, August 2003), p. 12. However, the "total number of elements" presented in the figure, "Summary of fault tree elements reviewed by the Board," only adds up to 1560. See p. 86.

   On p. 85, the report says, "Some of the fault trees are very large and intricate. For instance, the Orbiter fault tree, which only considers events on the Orbiter that could have led to the accident, includes 234 elements. In contrast, the Systems Integration fault tree, which deals with interactions among parts of the Shuttle, includes 295 unique multi-element integration faults, 128 Orbiter multi-element faults, and 221 connections to the other Shuttle components."

24. The System of War

1. W.E. Vesely, F.F. Goldberg, N.H. Roberts, and D.F. Haasl, *Fault Tree Handbook*, NUREG-0492 (Washington, DC: US Nuclear Regulatory Commission, January 1981), p. I-4.
2. W.E. Vesely, et al, *Fault Tree Handbook*, p. I-4.
3. W.E. Vesely, et al, *Fault Tree Handbook*, p. I-5.
4. Russell L. Ackoff. Noted by the author during Dr. Ackoff's keynote address at the 49th Annual Quality Congress, Cincinnati, Ohio, 22 May 1995.
5. For more on the latter, see the discussion of predatory conflicts in O'Connell, *Of Arms and Men*, e.g., p. 292.
6. I am grateful to LTC Tom Kratman for the suggestion; however, I will leave that exercise to others.
7. Recalling the discussions of the operational level from Chapter 12.
8. See Tom Kratman, "Indirectly Mistaken Decision Cycles."
9. According to the British Broadcasting Corporation's "e-cyclopedia," the term "exit strategy" is "a term from the venture-capital world referring to closing an investment" that was first applied to military situations by Secretary of Defense Caspar Weinberger in 1984. It was "popularised by General Colin Powell . . . in the early 1990s." "Exit strategy: How best to get out," in *E-CYCLOPEDIA: The words behind the headlines*. On-line, 20 October 2005, *http://news.bbc.co.uk/1/hi/magazine/3721983.stm*.
10. John Ringo and Tom Kratman, *Watch on the Rhine—Die Wacht am Rhein* (Riverdale, New York: Baen Books, 2005), p. 23.

25. The War Tree

1. W.E. Vesely, et al, *Fault Tree Handbook*, p. III-3.
2. W.E. Vesely, et al, *Fault Tree Handbook*, p. IV-1.

26. The Evolution of Theory

1. William Shakespeare, *Hamlet*, Act II, Scene 2. Polonius, referring to Hamlet.
2. Alger, *The Quest for Victory*, pp. 189-90.
3. See Chapter 23.

27. Elements of War

1. I owe considerable thanks to Professor James H. Galt-Brown, Abraham Baldwin Agricultural College, for challenging my assumptions and helping me refine my thoughts. Much of this section belongs to him.
2. The dehumanizing effect of war was discussed in Chapter 24. Several references to "predatory" conflict are throughout Robert O'Connell's *Of Arms and Men*.
3. As pointed out throughout O'Connell, *Of Arms and Men*.
4. See van Creveld, *The Transformation of War*.
5. Van Creveld covers this at length in *The Transformation of War*, but missed the mark a bit on p. 173 when he mentions "certainty": "Under actual historical circumstances, the principle factor affecting the question of certainty has not been perfect information, nor the lack of a credible defense, but the relationship of strength to weakness."
6. *Joint Publication 1*, 14 November 2000, p. V-3. The passage does not appear in the 14 May 2007 or later versions.

7. Resolution 687 (1991) formed a UN Special Commission to inspect and destroy Iraq's chemical, biological, and ballistic missile programs, and imposed a similar program, conducted by the International Atomic Energy Agency, of inspecting and destroying Iraq's nuclear weapons. See Note 7 to Chapter One for more details.
8. The two non-organizational definitions given are, "1. The authority that a commander in the Armed Forces lawfully exercises over subordinates by virtue of rank or assignment. 2. An order given by a commander; that is, the will of the commander expressed for the purpose of bringing about a particular action." *Joint Publication* 1-02, p. 43.

   Interestingly, the 2011 version of the DoD dictionary expanded the first definition as follows: "Command includes the authority and responsibility for effectively using available resources and for planning the employment of, organizing, directing, coordinating, and controlling military forces for the accomplishment of assigned missions. It also includes responsibility for health, welfare, morale, and discipline of assigned personnel."
9. Liddell Hart, *History of the Second World War*, p. 35.
10. I am grateful to Dr. James Galt-Brown for this observation.

    That the enemy "put both modern technology and modern political organization in the service of evil" meant that "To defend themselves from this threat, liberal democracies were led to adopt military strategies like the bombing of Dresden or Hiroshima that in earlier ages would have been called genocidal." Fukuyama, *The End of History*, p. 6.
11. Milton W. Meyer, *Japan: A Concise History*, Third Edition (Lanham, MD: Littlefield Adams Quality Paperbacks, 1993), pp. 76-7.
12. Meyer, *Japan: A Concise History*.

    The culture of the sword was ostensibly "broken" when Tokugawa samurai were defeated (just before the Meiji Restoration began in 1868) and then four more samurai revolts in 1873-77 culminated in the 1877 defeat of the Satsuma Rebellion. (O'Connell, *Of Arms and Men*, p. 236.) However, Japan's 20th Century warrior ethic had its roots in that samurai culture and tradition.

28. The Elements and the Principles

1. According to *Joint Publication* 1-02, "control" is "Authority that may be less than full command exercised by a commander over part of the activities of subordinate or other organizations" (p. 53).
2. *Joint Publication* 1, 12 July 2017, p. xxiii. See also pp. V-14 through V-18.
3. The AFC2ISRC name was itself not the first name of the unit in question, nor the last: the organization was realigned in 2010 as the Air Force C2 Integration Center (AFC2IC). The AFC2IC was deactivated three years later.
4. Sun Tzu, *The Art of War*, pp. 114-5.
5. Sun Tzu, *The Art of War*, p. 84.
6. Sun Tzu, *The Art of War*, chapter 11. Griffith names this chapter, "The Nine Varieties of Ground."
7. See e.g., William Arkin, "Planning an Iraqi War But Not an Outcome," *Los Angeles Times*, 5 May 2002, p. M-1, and Eric Schmitt, "U.S. Plan for Iraq is Said to Include Attack on 3 Sides," *New York Times*, 5 July 2002, p. A-1.
8. *Joint Publication* 5-0, *Joint Planning* (Washington, DC: US Department of Defense, 16 June 2017), p. I-4. On-line, 7 March 2018, *http://www.jcs.mil/Doctrine/Joint-Doctrine-Pubs/5-0-Planning-Series/*.

29. Theory Versus Theory

1. For an interesting re-reading of Weigley, see Antulio J. Echevarria II, *Toward an American Way of War* (Carlisle, PA: US Army War College Strategic Studies Institute, March 2004). For instance, on p. 1 Echevarria states, "the American approach to war was—to take the liberty of rephrasing Weigley's argument—more a way of *battle* than an actual way of war" (emphasis in original). Later (p. 10), he states, "the new American way of war seems headed for the same trap that snared . . . Weigley . . . , that is—it appears geared to fight wars as if they were battles and, thus, confuses the winning of campaigns or small-scale actions with the winning of wars."

   See also Major Michael Forsyth, USA, "Finesse: A Short Theory of War," *Military Review*, Volume 84, Number 4, July-August 2004, p. 18.
2. Charles Townshend, "The Shape of Modern War," in *The Oxford History of Modern War*, p. 19.
3. For one such attempt, see Eccles, *Military Concepts and Philosophy*, p. 28.

31. Case 1—Parity

1. Sun Tzu, *The Art of War*, p. 73.

32. Case 2—Single Advantage: Momentum

1. This was also covered in Chapter 19. See O'Connell, *Of Arms and Men.*
2. Keegan, *A History of Warfare*, p. 38.
3. See Liddell Hart, *Strategy*, p. 181.
4. See Sherry Sontag, Christopher Drew, and Annette Lawrence Drew, *Blind Man's Bluff: The Untold Story of American Submarine Espionage* (New York: Public Affairs, 1998).

33. Case 3—Single Advantage: Command

1. This idea is not original with the author; however, we cannot recall the source.
2. Sun Tzu, *The Art of War*, p. 87.
3. Sun Tzu, *The Art of War*, p. 96.
4. See McNeilly, *Sun Tzu and the Art of Modern Warfare*, pp. 5-6.
5. Keegan, *The Face of Battle*, p. 105 (footnote).
6. Keegan, *The Mask of Command*, especially p. 99.
7. See O'Connell, *Of Arms and Men.*
8. See Schwarzkopf, *It Doesn't Take a Hero.*
9. As pointed out by a number of sources, e.g., O'Connell, *Of Arms and Men.*

   For a counter argument, see Major Howard D. Belote, "The Weaponization of Space—It Doesn't Happen in a Vacuum," *Aerospace Power Journal*, Volume XIV, Number 1, Spring 2000, pp. 46-52.

34. Case 4—Split Advantage

1. Keegan, *Intelligence in War*, pp. 14-5.
2. Sun Tzu, *The Art of War*, pp. 61-2. (The anecdote about Sun Pin is in the biographical preface to the book.)

3. Keegan drew this inference from the storming of Crete in May 1941. *Intelligence in War*, p. 324.
4. As pointed out by Hank Reinhardt, swordmaker extraordinaire.
5. Liddell Hart, *Strategy*, p. 166.
6. Keegan, *The Second World War*, p. 275.
7. Keegan, *Intelligence in War*, p. 324.
8. Martin Van Creveld, "Technology and War II: Postmodern War?," in Charles Townshend, editor, *The Oxford History of Modern War* (Oxford: Oxford University Press, 2000), pp. 356-7.
9. Douglas Porch, "Imperial Wars: From the Seven Years War to the First World War," in Charles Townshend, editor, *The Oxford History of Modern War* (Oxford: Oxford University Press, 2000), p. 107.
10. See, e.g., Travis D. Taylor and Bob Boan, *Alien Invasion: The Ultimate Survival Guide for the Ultimate Attack* (New York: Baen Books, 2011).
11. Liddell Hart, *Strategy*, p. 322.
12. See Weigley, *The American Way of War*.

35. Case 5—Double Advantage

1. B.H. Liddell Hart, *History of the Second World War*, p. 21.
2. Sun Tzu, *The Art of War*, p. 92.

36. Elemental Questions

1. Sun Tzu, *The Art of War*, p. 85.
2. As stressed throughout Sun Tzu's *The Art of War*, especially pp. 92, 98, 134, and 136-7.
3. Wang Xi, in Sun Tzu, *The Art of War* (Cleary translation), p. 153.
4. General Sir Rupert Smith presents an excellent set of questions to ask "to define the outcome" of a military action "and the effort to be set to achieving it." *The Utility of Force*, pp. 384ff.
5. Sun Tzu, *The Art of War*, p. 84. See also p. 129.

37. The Search for Natural Laws of War

1. Liddell Hart, *History of the Second World War*, pp. 20 and 70.
2. See Liddell Hart, *Strategy*.

   Sun Tzu also wrote of the indirect approach. See *The Art of War*, pp. 102, 106.
3. McNeilly, *Sun Tzu and the Art of Modern Warfare*, p. 9.
4. McNeilly continues, ". . .if a commander is ignorant, does not understand or (worse) ignores these principles, he does so at his peril." *Sun Tzu and the Art of Modern Warfare*, p. 9.
5. Often pointed out by Dr. W. Edwards Deming in his seminars on continual quality improvement.
6. "Evolution" is used because it has become the colloquial term, despite the preferred name "natural selection."

   With respect to natural selection, the mutation of alleles and appearance of new species have been observed in the natural world, though I am unaware if those mutations have resulted in hardier species except perhaps in insects and bacteria—especially in those cases where outside environmental influences, e.g., insecticides and antibiotics, have killed off less hardy individuals. Whether this micro-evolutionary selection was "natural" would seem to be a matter of debate; and, since natural selection is not subject to experimentation to the same degree as physics, it may be many years before evolution, and especially macroevolution, may move beyond the theory phase. The time scale of natural selection is unfortunately, not conducive to human observation.

Evolution, then, is an excellent example of why our knowledge cannot encompass the whole universe. "The total picture of reality can be drawn only when the last experience of the last man, and the final facts of science, are summed up." Lewis, *Mind and the World Order*, pp. 16-7.

7. Lewis, *Mind and the World Order*, p. 343.

38. The Risk of War, Revisited

1. "The sorry state of peace, for both sides, that has followed most wars can be traced to the fact that, unlike strategy, the realm of grand strategy is for the most part *terra incognita*—still awaiting exploration, and understanding." Liddell-Hart, *Strategy*, p. 322.

   For an example of the term "grand strategy" applied to a level more akin to "strategy" *per se*, see Schwarzkopf, *It Doesn't Take a Hero*, p. 219.
2. Sun Tzu, *The Art of War*, p. 63.
3. See John T. Correll, "The Military-Civilian Gap," *Air Force Magazine*, Volume 83, Number 1, January 2000, p. 2.

   The downward trend may not be as severe as it once was: In recent elections, it seems, more separated and retired military members than usual have been elected to state and Federal offices.
4. John Keegan suggests that the military, especially because of the use of monetary inducements (e.g., pilot and other specialty bonuses, re-enlistment bonuses, scholarships), may be thought of as a mercenary class that "may take on some of the attributes of a profession." (*A History of Warfare*, p. 228.) Following this logic, doctors and other professionals may also be thought of as mercenaries rather than professionals; in some ways, and especially to cynics, that may be true.

   With respect to civilian leaders who may take up the sword more quickly than military leaders would like, General Schwarzkopf discussed in *It Doesn't Take a Hero* the pressures US Central Command felt from civilian leaders during the First Gulf War (see pp. 361-2, and p. 413).

   At the highest levels, this may not be as much of a concern as some think. Great military leaders have not always made good Presidents, and great Presidents—even great wartime Presidents—have often had little or no military experience.
5. Gray, *Modern Strategy*, p. 60.
6. The practice of hiding information inside other information, e.g., hiding text in the digital code of image files.
7. Lewis, *Mind and the World Order*, p. 179.
8. Clausewitz, *On War*, p. 157.
9. Clausewitz, *On War*, p. 86.
10. Perhaps, rather than looking for a universal theory, we should look for a multiversal one: maybe the complexity of the interactions themselves is paramount, rather than any supposed single explanatory theory.
11. Sun Tzu, *The Art of War*, p. 100.
12. Sun Tzu, *The Art of War*, p. 70.
13. Clausewitz, *On War*, p. 86.

    Previously (p. 85), Clausewitz asserted, "No other human activity is so continuously or universally bound up with chance."
14. Warden, *The Air Campaign*, p. 1.

    Colonel Warden adds on p. 143, "Of all mankind's activities, war is the most baffling and intriguing."

15. Eccles, *Military Concepts and Philosophy*, p. 27. (Emphasis in original.)
16. Jomini, *The Art of War*, p. 185.
17. Gray, *Modern Strategy*, p. 62.
18. Martin van Creveld, *The Transformation of War*, p. 116.
19. Sun Tzu, *The Art of War*, p. 142.
20. Keegan, *Intelligence in War*, p. 348.

ACKNOWLEDGMENTS

Who am I, a retired Air Force officer, engineer, and writer with an honorable enough but undistinguished service record and no combat experience, to think that I could categorize the elements of war and warfare in any useful way? New and even innovative, perhaps. Imaginative and audacious, certainly. Provocative and controversial, possibly. But useful? I must have been mad to attempt it.

Time has already told much in this work's development. As I stated in the Preface, this book changed radically from its initial inception until now. It could have continued to grow, as the sum total of military thought continues to grow, but never would I have caught up with that growth because for every book I could read—and there have been far too few of them—there are at least twenty more published that I should or would want to read. But at some point we have to "Cry, 'Havoc!' and let slip the dogs of war," so to speak.

Throughout the process of building this book, many people provided helpful insights and comments that influenced the assumptions, thoughts, and conclusions. I especially thank Colonel William Perkins, Colonel Michael Mantz, Lieutenant Colonel Tom Kratman, Lieutenant Colonel Deborah Fort, Captain Michael Scardera, Captain Mark Van Voorhis, and Doctor James H. Galt-Brown for interesting and enlightening discussions. Their careful reviews and challenging suggestions greatly improved the final product, and some of their contributions are credited in specific places in the text.

In the end, I can take full credit only for any errors that remain.

ABOUT THE AUTHOR

Lieutenant Colonel Graham W. "Gray" Rinehart retired from the United States Air Force in 2006 after a rather odd career. He kept rocket propulsion research operations safe, fought fires as head of a Disaster Response Force, trained Air Force ROTC cadets, refurbished space launch facilities, "flew" Milstar satellites, drove trucks, processed nuclear command and control orders as an Emergency Actions officer, commanded the Air Force's largest satellite tracking station, protected militarily critical space technologies, and wrote speeches for top Air Force leaders. He performed these duties at the Air Force Rocket Propulsion Laboratory, Edwards Air Force Base, California; Air Force ROTC Detachment 770, Clemson University, Clemson, South Carolina; the Titan System Program Office, Vandenberg Air Force Base, California; 4th Space Operations Squadron, Falcon Air Force Base, Colorado; 55th Mobile Command & Control Squadron, Offutt Air Force Base, Nebraska; Detachment 3, 22nd Space Operations Squadron, Thule Air Base, Greenland; the Defense Technology Security Administration, Alexandria, Virginia; and the Secretary and Chief of Staff of the Air Force Executive Action Group, Headquarters Air Force, the Pentagon.

Gray is the author of articles, essays, and papers that have appeared in such publications as *Air & Space Power Journal* and *Space Energy and Transportation*, as well as *Quality Education*, which was one of the first book-length treatments of industrial quality improvement philosophy applied to education. He has also written fiction: one novel, *Walking on the Sea of Clouds*, and short fiction in such venues as *Analog Science Fiction & Fact*, *Asimov's Science Fiction*, *Orson Scott Card's Intergalactic Medicine Show*, and various anthologies. He is currently a Contributing Editor for Baen Books, and a singer/songwriter with three albums whose music has appeared on *The Dr. Demento Show*.

Gray was married to the former Jill Marie Briggs of Murrells Inlet, South Carolina, whose untimely death in 2019 delayed release of this volume. He has two adult children, Stephanie and Christopher.

Made in the USA
Monee, IL
04 November 2026

67383563R00154